# Useful Optics

Chicago Lectures in Physics

Robert M. Wald, Editor
Henry J. Frisch
Gene F. Mazenko
Sidney R. Nagel

# Useful Optics

Walter T. Welford

The University of Chicago Press

Chicago and London

# Preface

The graduate student frequently needs to assemble some optical equipment as part of a research project. But the undergraduate training in optics in most physics degree courses is often inadequate for such a task. This is because of the sheer volume of other material that is considered an essential part of a modern physics degree.

While visiting the University of Chicago I was invited to give lectures to graduate students aimed at filling this gap; the present book is based on those lectures. Because my intention is to introduce the practical aspects of optical system development from the standpoint of graduate physicists, the coverage is somewhat different from that in most optics books. For example, it a reasonable assumption that electromagnetism courses will have familiarized the physics graduate with such things as polarization and the Fresnel formulas for reflection and refraction at dielectric interfaces. On the other hand, what forms the basis of a good deal of this book, Snell's law of refraction, particularly when it is expressed in three-dimensional vectorial form, might be less familiar. The inclusion of this book in a series of volumes of lecture notes is deliberate, for the book is intended to be used as the basis of or as a supplement to a lecture course and as a handy reference to have by the laboratory bench. Thus I have omitted proofs of formulas and I have included a small number of selected references where proofs and more detailed treatments can be found. However, the book is in no sense a textbook or monograph and there is no intention of complete coverage.

I am indebted to Sol Krasner, who first suggested that I write this book, to Robert M. Wald and Roland Winston for useful discussions, and to an unknown reviewer for very constructive criticism.

# 1

# Useful Models of Optics

The experimental physicist needs to collect light (or more generally electromagnetic radiation), to form images with it, and to manipulate and measure it. This book is intended as a practical aid to these purposes. Formulas will be given without proofs (but with adequate references), and the emphasis will be on applications.

For optics regarded as a tool, we shall use four models or physical approximations:

   (i) quantum optics,
   (ii) electromagnetic waves,
   (iii) scalar waves, and
   (iv) geometrical optics.

Of these, the quantum optics model will be needed only for discussions of the detection of radiation and of the noise present in detection. Electromagnetic wave theory is used in treating reflection and transmission at surfaces and multilayers, laser modes, and certain scattering problems. Scalar wave theory is helpful in explaining the bulk of diffraction and scattering problems and interference. Finally, geometrical optics is useful in a first approximation of light collection and image formation by almost any kind of optical system.

# 2

# Geometrical Optics

Because of its usefulness, we start with the geometrical optics model. In this model a "point source" emits rays which are straight lines in a vacuum or in a homogeneous isotropic dielectric medium. Light travels at different speeds in different dielectrics. Its speed is given by $c/n$, where $c$ is the speed in vacuum (299, 792, 458 m s$^{-1}$) and $n$, the *refractive index*, depends on the medium and on the frequency of the light.

A ray is refracted at an interface between two media. If $r$ and $r'$ are unit vectors along the incident and refracted directions, $n$ and $n'$ are the respective refractive indices, and $\mathbf{n}$ is the unit normal to the interface, then the ray directions are related by

$$n\mathbf{n} \times r = n'\mathbf{n} \times r', \qquad (2.1)$$

which is the law of refraction, *Snell's law*, in vector form. More conventionally, Snell's law can be written

$$n \sin I = n' \sin I', \qquad (2.2)$$

where $I$ and $I'$ are the two angles formed where the normal meets the interface, the angles of incidence and refraction. The two rays and the normal must be coplanar. Figure 2.1 illustrates these relationships and shows a reflected ray vector $r''$. Equation (2.1) can include this by means of the convention that after a reflection we set $n'$ equal to $-n$ so that, for reflection,

$$\mathbf{n} \times r = -\mathbf{n} \times r''. \qquad (2.3)$$

With a bundle or pencil of rays originating in a point source and traversing several different media, e.g., a system of lenses, we can measure along each ray the distance light would have traveled in a given time $t$; these points delineate a surface called a *geometrical wavefront*, or simply a *wavefront*. Wavefronts are surfaces orthogonal to rays (the Malus–Dupin theorem). (It must be stressed that wavefronts are a concept of geometrical optics and that they are *not* surfaces of constant phase (phasefronts) of the light waves in the scalar or electromagnetic wave approximations. However, in many situations the geometrical wavefronts are a very good approximation of phasefronts; see chap. 9.) Thus, if successive segments of a ray are of length $d_1$, $d_2$, ..., as in figure 2.2, a wavefront is a locus of constant $\Sigma\ nd$. This quantity is called an *optical path length*.

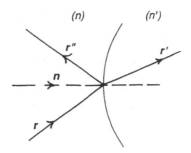

2.1 Snell's law in vector form, also showing the reflected ray.

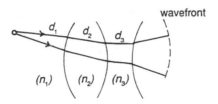

2.2 Generating a geometrical wavefront as a surface of constant optical path length from the source point.

Optical path lengths enter into an alternative to Snell's law as a basis for geometrical optics. Consider any path through a succession of media from, say, $P$ to $P'$. We can calculate the optical path length $W$ from $P$ to $P'$, and it will depend on the shape of this path, as shown in figure 2.3. Then *Fermat's principle* states that, if we have chosen a physically possible ray path, the optical path length along it will be stationary (in the sense of the calculus of variations) with respect to small changes of the path. (The principle as originally formulated by Fermat proposed a *minimum* time of travel of the light. Stationarity is strictly correct, and it means roughly that, for any small transverse displacement $\delta x$ of a point on the path, the change in optical path length is of order $\delta x^2$.) For our purposes, Fermat's principle and Snell's law are almost equivalent, but in the case of media of continuously varying refractive index it is sometimes necessary to invoke Fermat's principle to establish the ray path. Apart from such cases, either one can be derived from the other.

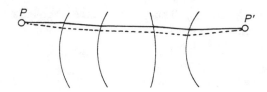

2.3 Fermat's principle. The solid line represents a physically possible ray between $P$ and $P'$, points which are not in general object and image, and the broken line a nearby path.

Either Fermat or Snell can be used to develop the whole edifice of geometrical optics in designing optical systems to form images.

# 3

# Symmetrical Optical Systems

The axially symmetric optical system, consisting of lenses and/or mirrors with revolution symmetry arranged on a common axis of symmetry, is used to form images. Its global properties are described in terms of *parazial* or *Gaussian* optics. In this approximation only rays making small angles with the axis of symmetry and at small distances from the axis are considered. The approximation is defined more precisely in chapter 6. In Gaussian optics, we know from symmetry that rays from any point on the axis on one side of the system emerge on the other side and meet at another point on the axis, the *image point*. This leads to the well-known formalism of principal planes and focal planes shown in figure 3.1. A ray entering parallel to the axis passes through $F'$, the second, or image-side, principal focus on emerging from the system, and a ray entering through $F$, the first principal focus, emerges parallel to the axis. A ray incident on the first, or object-side, principal plane $P$ at any height $h$ emerges from the image-side principal plane $P'$ at the same height $h$ so that the principal planes are *conjugate* planes of unit magnification. Excluding for the moment the special case in which a ray entering parallel to the axis also emerges parallel to the axis, these four points yield a useful graphical construction for objects and images, as depicted in figure 3.2.

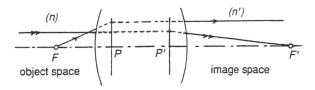

3.1 Definitions of principal foci, principal planes, etc. These do not necessarily occur in the order shown.

The two focal lengths $f$ and $f'$ are defined as

$$f \equiv PF, \qquad f' \equiv P'F'. \tag{3.1}$$

Their signs are taken according to the usual conventions of coordinate geometry, so that in figure 3.1 $f$ is negative and $f'$ is positive. The two focal

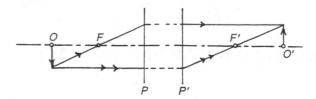

3.2 Graphical construction for object and image, leading to the conjugate distance equations.

lengths are related by

$$n'/f' = -n/f, \tag{3.2}$$

where $n$ and $n'$ are the refractive indices of the object and image spaces, respectively.

*Conjugate distances* measured from the principal planes are denoted by $l$ and $l'$, and the conjugate distance equation relating object and image positions is

$$n'/l' - n/l = n'/f' = -n/f. \tag{3.3}$$

The quantity on the right—that is, the quantity on either side of equation (3.2)—is called the *power* of the system, and is denoted by K.

Another form of the conjugate distance equation relates distances from the respective principal foci, $z$ and $z'$:

$$zz' = ff'. \tag{3.4}$$

This equation yields expressions for the transverse magnification:

$$\eta'/\eta = -f/z = -z'/f'. \tag{3.5}$$

It is useful to indicate paraxial rays from an axial object point $O$ to the corresponding image point $O'$ as in figure 3.3 with convergence angles $u$ and $u'$ positive and negative, respectively, as drawn in the figure. (Paraxial angles are small [see chap 6] but diagrams like figure 3.3 can be drawn with an enlarged transverse scale. That is, convergence angles and intersection heights such as $h$ can all be scaled up by the same factor without affecting the validity of paraxial calculations.) Then, if $\eta$ and $\eta'$ are corresponding object and image sizes at these conjugates, the following relation exists between them:

$$nu\eta = n'u'\eta'. \tag{3.6}$$

In fact, for a given paraxial ray starting from $O$, this quantity is the same at any intermediate space in the optical system. That is, it is an invariant, called the Lagrange invariant. It has the important property that its square is a measure of the light flux collected by the system from an object of size $\eta$ in a cone of convergence angle $u$.

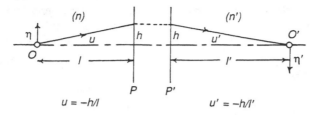

$$u = -h/l \qquad\qquad u' = -h/l'$$

3.3 Transverse magnification and the Lagrange invariant. The convergence angles $u$ and $u'$ are strictly paraxial quantities. The diagram is drawn according to the recognized convention, with greatly enlarged vertical scale.

The above discussion covers all general Gaussian optic properties of symmetrical optical systems. We next look at particular systems in detail. To do this, we abandon the skeleton representation of the system by its principal planes and foci and consider it as made up of individual refracting or reflecting surfaces.

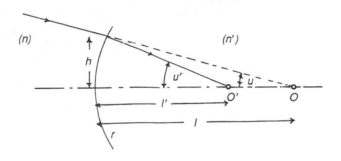

$$\frac{n'}{l'} - \frac{n}{l} = \frac{n'-n}{r} = K$$

$$n'u' - nu = -hK$$

3.4 The conjugate distance equation for a single refracting surface.

Figures 3.4 and 3.5 show the basic properties of a single spherical refracting surface of radius of curvature $r$ and of a spherical mirror. In each case $r$ as drawn is positive. These diagrams suggest that the properties of more complex systems consisting of more than one surface can be found by tracing paraxial rays rather than by finding the principal planes and foci, and this is what is done in practice. Figure 3.6 shows this with an iterative scheme outlined in terms of the convergence angles. The results can then be used to calculate the positions of the principal planes and foci (see, e.g., Welford 1986) and as the basis of aberration calculations.

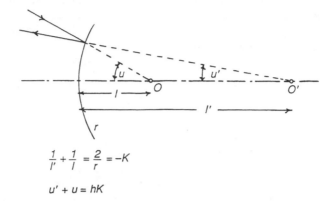

$$\frac{1}{l'} + \frac{1}{l} = \frac{2}{r} = -K$$

$$u' + u = hK$$

3.5 The conjugate distance equation for a single reflecting surface. By convention the refractive index changes sign after a reflection.

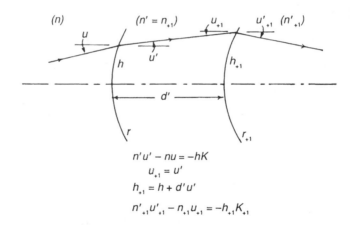

$$n'u' - nu = -hK$$
$$u_{+1} = u'$$
$$h_{+1} = h + d'u'$$
$$n'_{+1}u'_{+1} - n_{+1}u_{+1} = -h_{+1}K_{+1}$$

3.6 Paraxial raytracing. The equations shown are one possible set of many equivalent iterative schemes available according to personal choice.

The actual convergence angles which can be admitted, as distinguished from notional paraxial angles, are determined either by the rims of individual components or by stops deliberately inserted at places along the axis chosen on the basis of aberration theory. Figure 3.7 shows an *aperture stop* in an intermediate space of a system. The components of the system to the left of the stop form an image (generally virtual) which is "seen" from the object position (this image is usually virtual, i.e., it is not physically accessible to be caught on a ground-glass screen like the image in an ordinary looking glass); this image is called the entrance pupil, and it limits the angle of beams that can be taken in from the object. Similarly on the image side there is an exit pupil, the image of the stop by the components to the right, again usually virtual. These pupils may also determine the angles of beams from off-axis object points $O$ and $O'$; the central ray of the beam from $O$ passes through the center of the entrance pupil (and therefore through the center of the aperture stop and the center of the exit pupil) and it is usually called the *principal, chief* or *reference ray* from this object point. The rest of the beam or pencil from $O$ may be bounded by the rim of the entrance pupil, or it may happen that part of it is *vignetted* by the rim of one of the components.

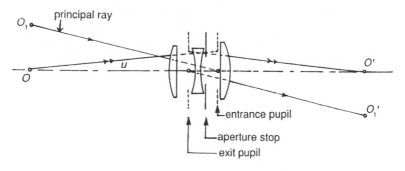

3.7 The aperture stop, the pupils, and the principal ray.

Although the aperture stop is usually thought of as being inside an optical system, as in a photographic objective, it is sometimes placed outside, and one example is the *telecentric stop* shown in figure 3.8. The stop is at the object-side principal focus, with the result that in the image space all the principal rays emerge parallel to the optical axis. A telecentric stop can be at either the object-side or the image-side principal focus, and the image conjugates can be anywhere along the axis. The effect is that the pupil on the opposite side of the telecentric stop is at infinity, a useful arrangement for many purposes. It may happen that the telecentric stop is between some of the components of the system.

The above information is all that is needed to determine how a given symmetrical optical system behaves in Gaussian approximation for any chosen object plane. Suitable groups of rays can be used to set out the system for mechanical mounting, clearances, etc. However, it is often easier and adequate in terms of performance to work with the *thin-lens model* of

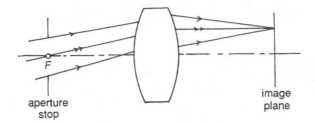

aperture
stop

image
plane

3.8 A telecentric aperture stop; the exit pupil is then at infinity. It is possible to have either the entrance pupil or the exit pupil or even both pupils at infinity. In the last case the principal ray would be parallel to the optical axis in both image and object spaces.

Gaussian optics. This model uses complete lenses of negligible thickness instead of individual surfaces. Figure 3.9 shows the properties of a thin lens. A system of thin lenses can be raytraced to find its properties, locate foci and ray clearances, etc., and very often the results will be good enough to use without further refinement. This is particularly true of systems involving unexpanded laser beams, where the beam diameters are quite small.

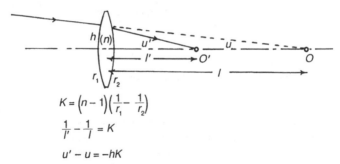

$$K = (n-1)\left(\frac{1}{r_1} - \frac{1}{r_2}\right)$$

$$\frac{1}{l'} - \frac{1}{l} = K$$

$$u' - u = -hK$$

3.9 The conjugate distance equation for a thin lens. The lens is drawn with a finite thickness, but its thickness is ignored in calculating its Gaussian properties. Thus the two principal planes coincide at the lens.

We omitted from our discussion of figure 3.1 the special case in which a ray incident parallel to the optical axis emerges parallel to the axis, as in figure 3.10. This is an *afocal* or *telescopic* system; it forms an image at infinity of an object at infinity, and the angular magnification is given by the ratio of the ray incidence heights. An afocal system also forms images of objects at finite distances, as indicated by the rays drawn in the figure. The transverse magnification is then constant for all pairs of conjugates. A good example of an afocal system is a laser beam expander (see figure 9.2).

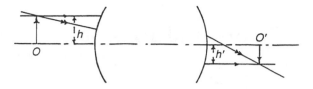

3.10 An afocal or telescopic system, shown with *finite* conjugates.

In many optical system designs, an effort must be made to get as much light through the system as possible, i.e., to increase signal strength, and this means using as large a diameter of aperture stop as possible. To a great extent the amount of light that gets through is governed by aberrations, departures of the rays from the simple paraxial description given in this chapter. We therefore defer discussion of this topic to chapter 6.

# 4

# Plane Mirrors and Prisms

A single-plane mirror used to deflect or rotate an optical axis needs no explanation, but some useful points can be made about combinations of mirrors. Two mirrors at an angle $\theta$ turn the beam through $2\theta$ about the line of intersection of the mirror planes whatever the angle of incidence on the first mirror, as in figure 4.1. The diagram is drawn for a ray in the plane perpendicular to the line of intersection of the mirror planes, but it is equally valid if the ray is not in this plane, i.e., the diagram is a true projection. In particular, if the mirrors are at right angles, as in figure 4.2, the direction of the ray is reversed in the plane of the diagram. Three plane mirrors at right angles to each other, forming a corner of a cube as in figure 4.3, reverse the direction of a ray incident in *any* direction if the ray meets all three mirrors in any order.

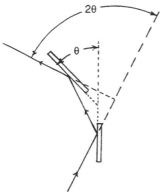

4.1 A pair of mirrors rigidly fixed together. The ray shown is in the plane containing normals to both mirrors, and whatever its angle of incidence, it is always turned through twice the angle between the normals.

These properties are more often used in prisms in the corresponding geometry. Total internal reflection, as in, for example, the right-angle prism (figure 4.4), is a great advantage in using prisms for turning beams. The condition for total internal reflection is

$$\sin I > 1/n. \tag{4.1}$$

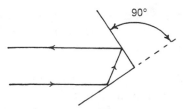

4.2 The roof reflector. The normals to the two mirror surfaces are at right angles so that a ray incident in a plane containing the normals is returned parallel to its original direction.

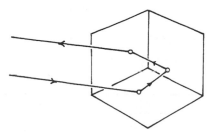

4.3 The cube-corner reflector. Each mirror is at right angles to the other two, and any ray which meets all three mirrors in turn is reflected back parallel to the original direction.

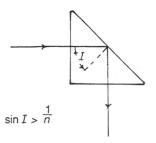

$$\sin I > \frac{1}{n}$$

4.4 The right-angle prism. All ordinary optical glasses fulfill the condition for total internal reflection at $45^\circ$ at visible wavelengths.

The *critical angle* given by $\sin I = 1/n$ is less than $45^\circ$ for all optical glasses, and probably for all transparent solids in the visible spectrum. Total internal reflection is 100% efficient provided the reflecting surface is clean and free from defects, whereas it is difficult to get a metallized mirror surface that is better than about 92% efficient. Thus with good antireflection coating on the input and output surfaces a prism such as that shown in figure 4.4 transmits more light than a mirror.

Roof prisms and cube-corner prisms, the analogues of figures 4.2 and 4.3, have many uses. The angle tolerances for the right angles can be very tight. For example, roof edges form part of the reversing prism system in some modern binoculars, and an error $\epsilon$ in the right angle causes an image doubling in angle of $4n\epsilon$. The two images are those formed by the portions of the beam incident at either of the two surfaces, which should have been at exactly $90°$.

In addition to turning the axis of a system, mirror and prism assemblies sometimes rotate the image in unexpected ways. The effect can be anticipated by tracing, say, three rays from a notional object such as the letter F (i.e., an object with no symmetry). A more direct and graphic method is to use a strip of card and mark arrows on each end as in figure 4.5a. The card is then folded without distorting it as in figure 4.5b to represent, say, reflection at the hypotenuse of the right-angle prism, and the arrows show the image rotation. The process is repeated in the other section, as in figure 4.5c. Provided the folding is done carefully, without distortion, this procedure gives all image rotations accurately for any number of successive reflections.

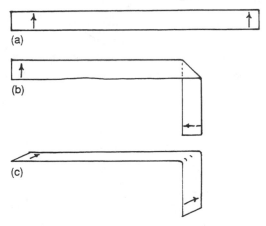

4.5 The draughtsman's paper strip method of checking image inversions. (a) The strip is marked with arrows at each end pointing the same way. (b) Then for, say, the right-angle prism of figure 4.4 the strip is folded without distorting or tearing it to represent a beam of rays in a plane perpendicular to the plane of incidence on the hypotenuse surface; the directions of the arrows show the relation between object and image orientations. (c) Then the strip is folded to represent rays in the plane of incidence. This method can be used for more complicated reflections provided the folding is done without distorting the paper.

The Dove prism (figure 4.6) is an example of an image-rotating prism. When the prism is turned through an angle $\phi$ about the direction of the incident light, the image turns in the same direction through $2\phi$. A more elaborate prism with the same function is shown in figure 4.7. The air gap indicated between the hypotenuses of the two component parts needs to be only about 10 $\mu$m or so thick to ensure total internal reflection. Any prism or mirror assembly like this with an odd number of reflections will serve as an image rotator. Figure 4.7 illustrates an elegant advantage of prisms over mirrors: the system can be made compact by using the same optical surface both for reflection and for transmission.

4.6 The Dove prism for image rotation. If the prism is turned through $\theta$ about the optical axis, the image rotates through $2\theta$. If the reflection is "unfolded," it can be seen that the Dove prism is the equivalent of a plane-parallel plate inclined at a finite angle to the optical axis; it would therefore introduce astigmatism and coma on the axis unless used where the beams from the object are collimated ("star space").

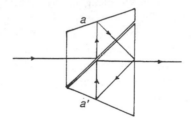

4.7 The Pechan prism for image rotation. The air gap must be at least about 10 $\mu$m wide to ensure total internal reflection of the obliquely incident beams. (Any system with an odd number of reflections acts as an image rotator.)

Figure 4.8 shows a typical beam-splitting (or combining) prism, a component of many diverse optical systems. The beam-splitting surface may be approximately neutral, in which case it would be a thin metal layer, or it may be dichroic (reflecting part of the spectrum and transmitting the rest), or it may be polarizing (transmitting the p-polarization and reflecting the s-polarization of a certain wavelength range). In the last two cases the reflecting-transmitting surface is a dielectric multilayer and its performance is fairly sensitive to the angle of incidence. Chapter 10 on multilayers deals with these properties in more detail. More on prisms for turning and reflecting beams is given by Smith (1966).

Prisms as devices for producing a spectrum have been largely replaced by diffraction gratings. The latter have several advantages for direct spectroscopy, but there are a few specialized areas where prisms are better. Losses in gratings through diffraction to unwanted orders are a nuisance

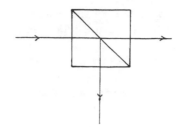

4.8 A beam-splitting cube. The reflecting surface can have various properties: neutral, i.e., equal reflecting and transmitting for all wavelengths; polarizing (the $p$-polarization is transmitted and the $s$-polarization is reflected); or dichroic, i.e., splitting the spectrum.

in certain astronomical applications where every photon counts. Another example of an area where prisms are preferable is wavelength selection in multiwavelength lasers: a prism inside the laser resonator with adequate angular dispersion ensures that only one wavelength will be produced, and one scans through the available wavelengths by rotating the prism. Figure 4.9 summarizes the notation for properties of dispersing prisms at and away from the position of minimum deviation. The significance of the minimum deviation position is that the effects of vibrations and placement errors are least. Also, if the shape of the prism is isosceles, the resolving power will be a maximum at minimum deviation. The main formulas relating to dispersing prisms are as follows.

Spectroscopic resolving power:

$$\lambda/\Delta\lambda = (t_1 - t_2)dn/d\lambda, \qquad (4.2)$$

where $t_1 - t_2$ is the difference between the path lengths in glass from one side of the beam to the other.

Angular dispersion:

$$dI_2'/d\lambda = \frac{\sin A}{\cos I_1' \cos I_2'} \cdot dn/d\lambda \qquad (4.3)$$

$$= \frac{2\sin(A/2)}{\cos I_1} \cdot dn/d\lambda \qquad (4.4)$$

at minimum deviation.

Spectrum line curvature:

$$1/\text{radius} = \frac{n^2 - 1}{nf} \cdot \frac{\sin A}{\cos I_1' \cos I_2'} \qquad (4.5)$$

$$= \frac{n^2 - 1}{n^2 f} \cdot 2 \cdot \tan I_1 \qquad (4.6)$$

at minimum deviation, where $f$ is the focal length of the lens which brings the spectrum to a focus.

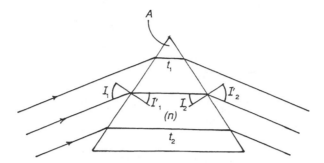

4.9 Notation for the properties of dispersing prisms.

The spectrum line curvature refers to the observation that the image of the entrance slit of the spectroscope produced by placing a lens after the prism is actually parabolic. The parabola is convex toward the longer wavelengths. The reason the image is curved is that rays out of the principal plane of the prism are deviated more than rays in the plane, a straightforward consequence of the application of Snell's law. For rays with angle $\epsilon$ out of the plane the extra deviation can be parametrized by an additional contribution to the index of refraction given by

$$dn \approx \epsilon^2(n^2 - 1)/(2n). \tag{4.7}$$

If the length of the slit image is $L$, then $\epsilon \approx L/(2f)$, where $f$ is the focal length of the lens. Moreover, from equation (4.3) we have

$$dI_2'/dn = \frac{\sin A}{\cos I_1' \, \cos I_2'} \tag{4.8}$$

$$= (2/n)\tan I_1 \tag{4.9}$$

at minimum deviation. The curvature of the slit image readily follows from these relations.

The typical dispersing prism of constant deviation shown in figure 4.10 has the property that, if it is placed in a collimated beam, the wavelength which emerges at right angles to the incident beam is always at minimum deviation so that the spectrum is scanned by rotating the prism about a suitable axis such as the one indicated.

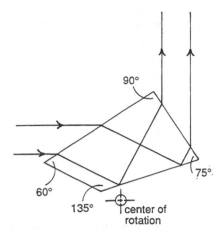

4.10 The Pellin-Broca constant-deviation dispersing prism. The center of rotation is chosen to keep the transmitted beam nearly undisplaced as the prism rotates to scan the spectrum. The wavelength, which is turned through $90^{\circ}$, is transmitted at minimum deviation.

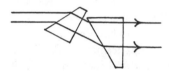

4.11 A pair of prisms used as a one-dimensional beam expander for a collimated beam.

A prism used a long way from minimum deviation will expand or contract a collimated beam in one dimension. Figure 4.11 shows a pair of prisms used in this way to turn a laser beam of elliptical profile (from a diode laser) into a beam of circular profile by expanding it in the plane of the diagram only.

# 5

# Optical Materials

It is convenient to classify optical materials into four groups: transparent materials for the ultraviolet, for the visible and for the infrared, and materials for mirrors. The topic of optical materials is to some extent connected with the subject of optical tolerances, to be discussed in chapter 7, but here we can note that such questions as homogeneity of refractive index, and the achievable accuracy of surface polish are often governed by the rule that the effect on a phasefront of correct shape should not be more than a quarter-wavelength (the Rayleigh $\lambda/4$ rule) for a reasonably high quality optical system; this gives tolerances based on the variation of optical path length (refractive index) integrated along a ray path due to inhomogeneity. An alternative set of specifications based on angular deviation tolerances through a supposedly plane-parallel slab of material leads to a refractive index gradient tolerance.

## 5.1 Optical Materials for the Visible Region

For high-quality optical systems *optical glasses* are used, glasses prepared with careful attention to uniformity of composition and with a range of optical properties. The main optical properties of such glasses are refractive index and chromatic dispersion. The dispersion is generally specified by the so-called *V*-value, defined by

$$V = (n-1)/\Delta n, \tag{5.1}$$

where $\Delta n$ is the difference in refractive index between two specified wavelengths, usually of chosen spectrum lines. Specialist optical glass manufacturers summarize this information on a chart such as that shown as figure 5.1, in which points indicate a glass of the refractive index and *V*-value corresponding to the coordinates. The outlined area indicates roughly the boundary of the region of obtainable glasses.

The manufacturers' catalogs also give more detailed information in tabular form for each glass: the refractive index at wavelengths from the near-ultraviolet to the near-infrared, including commonly used laser wavelengths, the transmission as a function of wavelength, and other properties which

19

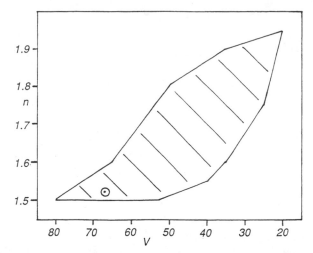

5.1 The glass chart. This is a quick-reference display of types of optical glass, as used by all manufacturers. A glass type is indicated by a point giving the refractive index for the middle of the visible spectrum and the dispersion ($V$-value). The hatched area indicates the region of available glass types.

may or may not be relevant to a particular application. We list below some of these properties together with others, indicated by an asterisk, which are not always given in current optical glass catalogs.

Thermal expansion

Dependence of refractive index on temperature

Nonlinear refractive index (dependence on square and higher powers of electric field strength)*

Verdet constant (for Faraday rotation)

Density

Elastic moduli

Stress birefringence (mechanical and electrical)*

Transformation temperature (for annealing)

Resistance to chemical attack

Refractive index inhomogeneity (absolute change is given but not the gradient)

Frequency of bubbles and inclusions

Light scattering*

Thermal expansion and index change with temperature both change the Gaussian properties of a system. Thus the power $K$ of a thin lens depends on temperature $T$ as follows:

$$dK/dT = \{(dn/dT)/(n-1) - \alpha\}K, \qquad (5.2)$$

where $\alpha$ is the thermal expansion coefficient.

Such effects matter when the system cannot be refocused to allow for large temperature changes, for example, optical systems in satellites or aircraft. It is then possible to design temperature-compensated systems by choosing suitable materials for lens mounts, spacers, etc. The expansion coefficients of optical glasses are of the order of magnitude of 4 to $7 \times 10^{-6}$ per degree Celsius. The change of refractive index relative to air with temperature is of order $10^{-6}$ per degree Celsius, with strong dependence on wavelength and the kind of glass.

Nonlinear index changes occur in glasses when the electric field strength is high, which causes self-focusing and, eventually, breakdown of the glass. It is to be watched for in pulsed laser systems and can even occur near caustics formed near focal regions in multiply reflected beams from nominally antireflection coated surfaces.

Faraday rotation is used in many contexts, for instance, in nonreturn switches to stop reflected laser pulses from reentering a laser resonator.

The quality of annealing of optical glass is so good nowadays that it is very unusual to find any serious stress birefringence from inadequate annealing. But for systems to be used for polarization measurements the manufacturers will supply specially annealed material. On the other hand, stress birefringence is sometimes deliberately used in systems, for example, in variable polarization retarders.

Many thin-film coating processes involve heating a glass substrate. In such processes the transformation temperature is a warning point. It is generally regarded as best not to heat a glass substrate within $100°\,C$ of this point to avoid the risk of irreversible change of shape and surface quality.

Much detail is available in the catalogs about chemical attack. If an optical system is to be used in a hostile environment, say, the tropics or a chemistry laboratory, it is advisable to seal it carefully and make all external optical surfaces of resistant glass, such as one of the borosilicate crowns.

Some glasses can be obtained in useful sizes with a refractive index checked as uniform throughout the piece to $\pm0.000001$. However, the tolerancing should be done carefully to determine if this degree of uniformity is really necessary; one pays heavily in cost and delivery time for such a high specification.

The choice from among perhaps 200 optical glass types in the manufacturers' catalogs is often made on the basis of refractive index and dispersion, as determined by details of the aberrational design. Sometimes, however, these details are not very relevant—for example, in a laser system where there is no question of chromatic aberration correction but where nonlinear effects may matter, or in a pressure window through which precision measurements are to be made. It is in such cases that attention must be paid to many nonoptical properties of optical glasses.

Plastics as optical materials have been in use for many years. Recently, improved molding techniques have made them even more popular. The range of available refractive indices and dispersions is very limited compared to that for optical glasses (refractive index 1.49 to 1.6 and $V$-value for the same wavelength range as in figure 5.1, i.e., 57 to 30); also, the dependence of index on temperature and the thermal expansion coefficient are both at least an order of magnitude greater than for optical glass. Thus plastics should be considered only for certain special cases where these disadvantages do not matter and where the high cost of molds is justified by large volume production. In spite of improvements in molding techniques it is still true that the homogeneity and accuracy of the surface shape of glass components cannot be approached by plastics.

## 5.2 Optical Materials for the Ultraviolet

For the present discussion the ultraviolet extends from about 170 nm to 400 nm; the lower limit is the limit of transmission by air in the laboratory. No production optical glasses transmit below 350 nm* and the range of available materials is very limited. In fact the only glassy (noncrystalline) material available is fused silica ($SiO_2$). Depending on the grade this material can transmit to about 180 nm. It is very hard and can be polished well to produce components of the highest quality. Also, it has a very low thermal expansion coefficient, $0.5 \times 10^{-6}$, but a surprisingly high temperature coefficient of refractive index, $8 \times 10^{-6}$. Manufacturers offer various grades of homogeneity and uv transmission.

Other uv transmitting materials are crystalline. Those which are cubic, i.e., optically isotropic and therefore not birefringent, are mainly the alkali halides, and of these the best for transmission, hardness, and workability seems to be lithium fluoride. Nevertheless lithium flouride is greatly inferior to fused silica in workability since it easily chips and scratches and since it is difficult to get a good polish at the same time as a good surface shape. However, lithium fluoride and other alkali halides offer the possibility of making achromatic combinations with fused silica since the dispersions are

---

* This is true at the time of writing, but much development work is being devoted to lowering this limit.

very different.

Other noncubic crystals have restricted use as windows, notably crystal quartz, which has rotary polarization in addition to being birefringent, and sapphire ($Al_2O_3$), which transmits down to 140 nm but also is birefringent. Calcium carbonate, variously and confusingly known as calcspar, Iceland spar, and calcite, transmits to below 200 nm and because of its very high birefringence is used for polarizing prisms in the uv as well as the visible.

## 5.3 Optical Materials for the Infrared

"Infrared" here means the two wavelength bands which are transmitted by the atmosphere, from 3 to 5 $\mu$m and from 8 to 14 $\mu$m. The latter is also known as the *thermal band* because the maximum of the blackbody spectrum at 300 K is at 10 $\mu$m, in the middle of this band.

The most popular materials for the thermal band are germanium, zinc selenide, and zinc sulphide. Germanium has the very high refractive index of 4.0 and also very low dispersion, which makes it easier to get good aberration correction with optical systems of wide aperture and field angle. Its transmission range is from 1.8 to 23 $\mu$m; that is, it is opaque in the visible region. Zinc selenide and zinc sulphide are similar materials: both have refractive indices of about 2.3, much higher dispersions than germanium, and transmission ranges from 0.5 to 20 $\mu$m (ZnSe) and from 0.4 to 20 $\mu$m (ZnS); the partial transmission in the visible region is useful for inspection. All three materials are available in both poly- and monocrystalline (cubic) form.

A wider variety of materials is available for the 3 to 5 $\mu$m band, including certain "chalcogenide" glasses and crystals such as sapphire, rutile, and calcite.

For more on infrared materials see Wolfe and Zissis (1978), and for definitive tables of refractive index and absorption of a range of optical materials, excluding glasses, for wavelengths from the ultraviolet to the infrared see Palik (1985). Also, the manufacturers of special crystals often give reliable data.

## 5.4 Materials for Mirrors

For mirrors below, say, 100 mm diameter which are not to be used in high-intensity beams almost any glass is adequate, for example, float glass or Pyrex. If the shape or "figure" is very critical, it is better to use carefully annealed material since optical working may release deforming stresses. Very high quality surfaces such as Fabry-Perot mirrors, which have to be flat to about $\lambda/200$, are made of fused silica on account of its low expansion coefficient. Mirrors for very high power lasers, such as $CO_2$ lasers for metal machining, are often water-cooled and made of copper or other metals; to get a good optical finish and high reflectivity, a process of electrodeless nickel coating is used, as outlined in figure 5.2.

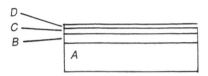

5.2 The electrodeless nickel process for mirrors on metal substrates. $A$, metal substrate, e.g., aluminum. $B$, coating of 90–10 nickel-phosphorus, 25 to 50 $\mu$m thick; this is hard enough to be optically worked to a good figure, but it has low reflectivity. $C$, evaporated aluminum for high reflectivity. $D$, protective coating of silica.

If a mirror is larger than about 100 mm diameter, it may be necessary to be more seriously concerned about temperature effects, depending on the accuracy of figure required. For astronomical telescope mirrors, which may be a few meters in diameter, the choices of both material and shape (thickness relative to diameter, whether ribbed, etc.) are critical. Two very similar materials were developed about 20 years ago specifically for large mirrors: Cer-Vit (Owens-Illinois Company) and Zerodur (Schott). They are slightly devitrified glasses, scattery in transmission, and light brown in color. By adjusting the heat treatment in the annealing process, they can be made to have an expansion coefficient of less than $10^{-7}$ per degree Centigrade, that is, less than one-fifth that of fused silica. Also, they take a good polish and seem to be generally ideal for making large mirrors. Their great dimensional stability (better than Invar, apparently) has led to their use as spacers for laser resonators and similar applications where stability is needed.

# 6

# Aberrations

In chapter 3 Gaussian or paraxial optics was introduced in order to get the first approximation of the optics of symmetrical optical systems. That this is an approximation follows from a simple trial which can be done either numerically or experimentally. Take a planoconvex lens as in figure 6.1 of focal length, say, twice its diameter. It is clear that from the use of Snell's law of refraction as described in section 2 we could, with the aid of some simple geometry, calculate the paths of rays through this lens without approximations, a procedure called *raytracing*. Figure 6.1 shows the result of doing this for a beam of rays coming from an object point at infinity on the axis. The rays close to the axis meet at the Gaussian image point, in this case the image-side principal focus, and rays further from the axis intersect the image plane away from the Gaussian image point. Figure 6.2 shows rays from infinity inclined at 10° to the axis (an off-axis object point); these are only rays in the plane of the diagram, and *skew* rays would complicate the pattern even further. Figures 6.1 and 6.2 illustrate only two examples of the great variety of ray aberrations, but they immediately suggest that aberrations depend on the diameter of the lens and on the distance of the object point from the axis. For off-axis object points and skew rays there must be a third parameter.

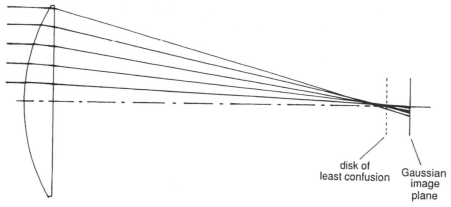

disk of
least confusion

Gaussian
image
plane

6.1 Raytrace showing spherical aberration.

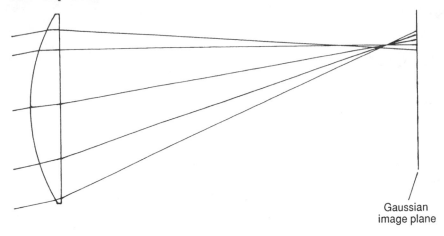

Gaussian
image plane

6.2 Raytrace showing off-axis aberrations in the tangential (meridian) section. The aberration shown is mainly coma.

These deviations from the Gaussian image point are known as transverse ray aberrations. It can be useful for several purposes to consider aberrations also in terms of the shapes of the geometrical wavefronts defined in chapter 2. Since wavefronts are orthogonal to rays, if all the rays meet at a single point the wavefronts in the beam must be portions of spheres. Conversely, if the wavefronts are not truly spherical, the rays will not meet at a point so that aberration in a beam can be measured as the departure of one of its wavefronts from spherical shape.* Thus the wavefront is compared with a spherical surface centered at the ideal or Gaussian image point and touching the wavefront at the center of the exit pupil. The optical path length from sphere to wavefront along a ray is then a measure of the aberration as a function of position of the ray in the pupil and position of the object point in the field.

We can parameterize aberrations in more detail in terms of three coordinates: $\rho$, $\phi$, and $\eta$; $\rho$ is the distance from the axis to the point at which a ray meets the entrance pupil, $\phi$ is its azimuth with respect to an origin in the plane of the diagram (figure 6.3), and $\eta$ is a *field coordinate*, that is, it measures in one of a few different ways how far off-axis the object point is. It is found that on account of the axial symmetry the wavefront aber-

---

* If one wavefront is not spherical none of the others will be, since they are all parallel surfaces, and the departure from spherical will change from wavefront to wavefront. In practice for a system forming even moderately good images this effect is small; usually the particular wavefront to which aberration is referred is that at the exit pupil.

rations depend on $\rho^2$, $\rho\eta\cos\phi$, and $\eta^2$ and can be expressed as a power series in those variables. The terms of lowest order in $\rho^2$, $\rho\eta\cos\phi$, and $\eta^2$ correspond to Gaussian optics, and those involving higher powers are various forms of aberration.

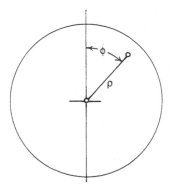

6.3 Polar coordinates in the exit pupil for parameterizing aberrations.

Thus Gaussian optics can be formally defined as optics in the region close to the axis such that terms higher than the square of $\rho$ and $\eta$ in expansions of optical path lengths can be ignored. From the discussion of optical tolerances (chap. 7) this definition can be sharpened to say that in the Gaussian optics region terms higher than squares of $\rho$ and $\eta$ must be much smaller than the wavelength of the light to be used.

Higher-order terms in the power series expansion represent aberrations. The next terms beyond paraxial are called primary, third-order, or Seidel aberrations; they are of fourth degree, and there are five of them as follows:

$$C_1\rho^4, \tag{6.1}$$

$$C_2\rho^3\eta\cos\phi, \tag{6.2}$$

$$C_3\rho^2\eta^2\cos^2\phi, \tag{6.3}$$

$$C_4\rho^2\eta^2, \tag{6.4}$$

$$C_5\rho\eta^3\cos\phi. \tag{6.5}$$

(There is, of course, a sixth term in the variables, namely $\eta^4$, but since this is constant over the pupil, it leaves the wavefront spherical in shape and is therefore not classed as an aberration in the present context.)

The magnitudes of the Seidel aberrations can be found either from formulas involving the constructional parameters of the system (curvatures, refractive indices, and separations) or by raytracing followed by polynomial fitting. For details of these processes see Welford (1986). Here we shall describe in detail the forms of these terms as both wavefront shapes and ray deviations, and we shall indicate how some aberrations can be controlled by appropriate optical design.

## 6.1 Spherical Aberration

The term in $\rho^4$, for historical reasons, is known as *spherical aberration*. Figure 6.4 shows how the effect on wavefront shape may be represented: an axial object point is assumed to form a paraxial image at $O'$, and if there were no aberration, a wavefront of this beam would be spherical with its center at $O'$. This ideal wavefront is called the reference sphere, and an actual wavefront with spherical aberration is indicated, coinciding with the reference sphere at the axial point. The rays would be the normals to the wavefront, and it can be seen that they would intersect the axis further to the left of the paraxial image point. In fact the ray pattern would be as in figure 6.1, which is drawn for spherical aberration. It is sometimes useful to plot the deviation of the rays from the paraxial image point in the image plane as a result of the aberration, as in figure 6.5, where the vertical coordinate is the ray height in the pupil and the horizontal coordinate is the distance in the image plane from the paraxial image point; this is a plot of transverse ray aberration. We could also plot the ray intersection along the axis against pupil height, the *longitudinal spherical aberration*. Finally, we could plot a pattern of dots in the image plane representing the ray intersections, as in figure 6.6; this is called a *spot diagram*. All of these representations of spherical aberration and corresponding representations of the other aberration types to be described have their uses in assessing the performance of a system.

Referring again to figure 6.1, it can be seen that, if the focal plane is moved slightly nearer to the lens, as indicated by the broken line, the aberration patch is smaller so that the Gaussian image plane is not necessarily where the sharpest image will be with spherical aberration. In the wavefront picture this is represented by taking a reference sphere centered on a new image point, as in figure 6.4, where the new reference sphere is shown in broken line. We see that in this case the maximum value of wavefront aberration is less than for the Gaussian focus and that the aberration changes sign at a certain radius in the pupil. In actually using an optical system with spherical aberration, one would usually be able to set the image at the best focus according to whatever criterion of the quality of the image

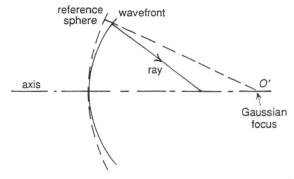

6.4 A wavefront with spherical aberration showing the reference sphere centered on the Gaussian image point.

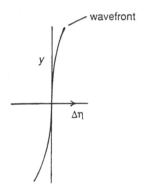

6.5 Transverse ray aberration in the case of spherical aberration.

was appropriate; the present discussion is intended to show how this might be allowed for in a numerical analysis of a proposed design.

Terms of higher order than the Seidel term are often significant for spherical aberration, e.g., $\rho^6, \rho^8, \dots$. To illustrate this, it is convenient to show the wavefront aberration for an image point at infinity, or in what is sometimes called star space. Figure 6.7 shows a combination of fourth and sixth power spherical aberration of opposing signs relative to a plane reference sphere. Here it may be seen again that a change in focus amounting to a slight curvature of the reference sphere may be beneficial.

```
                              SIMPLE LENS
    ├───────┤  2.0000mm    File              Defocus=%-10.0000mm  Inc=2.0000mm

      -14.0000        -12.0000       -10.0000       -8.0000        -6.0000
```

6.6 Spot diagrams for spherical aberration calculated for the lens of figure 6.1 at different planes of focus centered around the disk of least confusion. The distances in mm from the paraxial focus are indicated above each pattern.

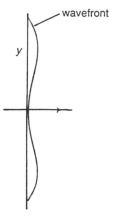

6.7 Primary and secondary spherical aberration as wavefront aberration (4th- and 6th-power terms).

## 6.2 Coma

We saw in section 6.1 that spherical aberration depends only on the aperture variable $\rho$; i.e., it is independent of the field variable $\eta$. There is a group of aberrations called *linear coma* which depend linearly on the field variable and, in the case of Seidel coma, on the cube of the aperture variable (eqn. 6.2); thus in star space the wavefront with linear coma is as in figure 6.8, or alternatively it can be represented as a contour map as in figure 6.9. To describe this we have to introduce more terminology: the section of the wavefront which contains the plane through the optical axis is called the *meridian section* (sometimes the *tangential section*), and the plane pependicular to it and passing through the principal ray defines the *sagittal section*. Equation (6.2) indicates a cubic dependence of the wavefront aberration on the aperture coordinate in the meridian section, as shown in figures 6.8 and 6.9. In the sagittal section the wavefront aberration is zero but the wavefront has a finite gradient in the circumferential direction. The cubic dependence in the tangential section and the gradient across the sagittal section both depend linearly on the field coordinate so that although the coma is zero on axis it increases rapidly in the field if the coma coefficient is nonzero. The ray pattern for coma is quite complex. Figure 6.10 shows a diagram drawn for rays distributed uniformly around circles of increasing diameter in the exit pupil. For each circle in the exit pupil the ray intersections trace out the corresponding circle in the figure twice. The marked asymmetry of the ray pattern illustrates that coma has a bad effect in systems where the positions of points in the image are to be determined accurately, e.g., in astrometry and surveying.

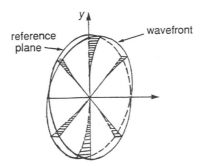

6.8 The wavefront shape for coma.

## 6.3 Astigmatism and Field Curvature

In the Gaussian optics approximation, rays from one object point all meet at one image point but if another image plane is chosen the rays meet it in a uniform patch, as in figure 6.11. Alternatively, we may regard such a shift of focus as a wavefront aberration depending on the square of the aperture coordinate, again as in figure 6.11. This may seem pointless in Gaussian optics, but it is a useful notion in discussing aberrations. *Field curvature* is an aberration in which the focal position changes according

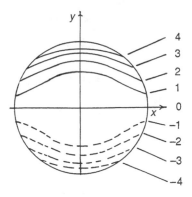

6.9 A contour map of the wavefront for coma. Each contour represents, say, an increment in height of the wavefront of $\lambda/2$.

6.10 Coma as transverse ray aberration. The circles are loci of intersections of rays from a point object through concentric circles in the pupil. Once around a circle in the pupil corresponds to twice around the corresponding circle in the image.

to the square of the field angle; thus on the axis the rays are in focus at the Gaussian image plane but the off-axis images are formed on a curved surface, as in figure 6.12, so that the wavefront aberration is represented by equation (6.4).

*Astigmatism* similarly depends on the square of the aperture coordinate, but it involves a change of shape of the wavefront from the ideal spherical shape. It appears as focal shifts in the meridian and sagittal sections, which are of different magnitudes, so that the wavefront aberration is as in equation (6.3). The effect is that the rays all pass through two lines in the focal region rather than a single point. These lines are known as the tangential and sagittal focal lines. Midway between the focal lines the rays pass through a circle called the disk of least confusion.

Both astigmatism and field curvature as Seidel aberrations depend on the square of the field angle so that a diagram such as that shown as figure 6.13 can be used to represent the effects of both. However, as with the other aberrations there are higher order effects in real systems which change this relatively simple picture.

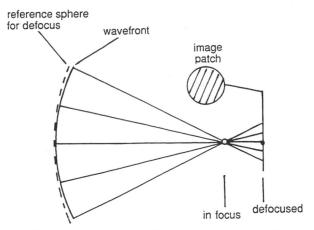

6.11 The effect of change of focus on an unaberrated beam.

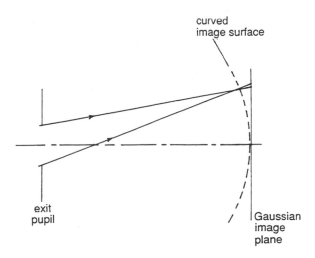

6.12 Field curvature.

## 6.4 Distortion

The last Seidel aberration, *distortion*, has, unlike most of the others, a self-explanatory name. The wavefront term, equation (6.5), means that the wavefront is not changed in shape but is simply tilted with respect to the reference sphere centered on the off-axis Gaussian image point to an extent depending on the cube of the field coordinate. This has the effect

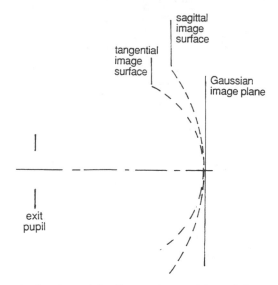

6.13 Astigmatism. Rays in the plane of the diagram (tangential rays) focus in a line on the tangential image surface; sagittal rays focus in a line on the sagittal image surface.

of changing the magnification according to the position in the field; thus a square centered in the object plane is imaged as shown in figure 6.14.

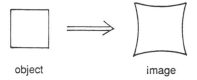

6.14 Distortion. An object in the form of a square is imaged as indicated in "pincushion" distortion. The opposite sign of aberration gives "barrel" distortion.

## 6.5 The Effect of the Position of the Aperture Stop

The aperture stop and the associated entrance and exit pupils were described in chapter 3 as having the function of limiting the extent of the beams of rays taken in by the optical system. From the discussion in the present chapter it can be seen that the size of the aperture stop also controls the amount of any residual spherical aberration that of necessity remains in the optical design (that there always are residual aberrations follows from limits on the skills of optical designers and on the permissible cost and complexity of optical systems).

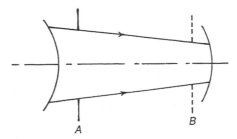

6.15 Changing the position of the aperture stop. When the stop is moved along the axis from *A* to *B* in the same air space, its diameter is changed so as to keep the size of the axial beam constant.

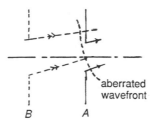

6.16 The effect on off-axis aberrations of moving the aperture stop. The change from *A* to *B* selects a different portion of the aberrated wavefront.

There is a third function of the aperture stop, the control of off-axis aberrations. Figure 6.15 shows the space in a system which contains an aperture stop, shown in full line. Clearly, the stop could be displaced along the axis, as indicated by the arrow, to the position shown in broken line, and provided its diameter is changed appropriately, there will be no change in the size of the beams collected and in the spherical aberration correction. To see the effect on off-axis aberrations, consider the off-axis wavefront shown in figure 6.16; the exit pupil in the position shown defines the principal ray and permits the indicated part of the aberrated wavefront to pass, but if the axial movement of the stop moves the exit pupil to the broken line position, a new principal ray and a different portion of wavefront are selected. In this way it can be seen that shifting the stop can change the off-axis aberrations. The extent to which this effect can be used in optical design is often limited by practical considerations, but it cannot be disregarded except in systems such as collimators which have negligible field of view.

## 6.6 Chromatic Aberrations

We discussed optical materials in some detail in chapter 5. Here we recall that the refractive index of all materials depends on the wavelength of the light used; generally for transparent materials the index decreases with increasing wavelength. Thus both Gaussian and aberrational properties of refracting optical systems generally depend on wavelength; i.e., there is *chromatic aberration*. In fact what is usually meant by chromatic aberrations are variations of Gaussian properties: (a) a shift of image-plane distance with wavelength, called longitudinal chromatic aberration or longitudinal color, and (b) a change of magnification—transverse chromatic aberration or transverse color. The higher-order effects are referred to by self-explanatory terms such as chromatic variation of spherical aberration, etc. Figure 6.17 shows the Gaussian chromatic effects; longitudinal color appears as a shift of the whole image plane along the axis, and transverse color is a radial shift of the image point proportional to the field coordinate.

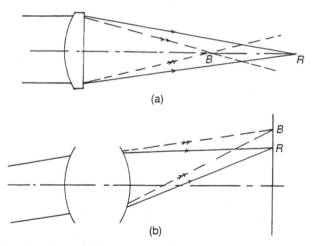

6.17 Chromatic aberration.  (a) Longitudinal chromatic aberration, in which different wavelengths focus at different distances along the axis. (b) Transverse chromatic aberration, in which the magnification varies with the wavelength.

## 6.7 Dependence of Aberrations on Optical System Design

The aberrational behavior of a system can be determined in as much detail as desired by tracing enough rays through it according to Snell's law. This is in principle a simple procedure although it does not always offer much insight as to why the system behaves as it does or how it might

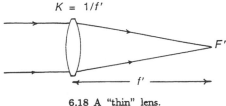

6.18 A "thin" lens.

be improved. Calculating the Seidel aberrations gives a more analytical insight, since the contributions of each refracting or reflecting surface can be determined. The equations for these calculations are given in books on aberration theory (see, e.g., Welford 1986); also, there are now plenty of software packages available for desktop computers which will do such calculations and present the results in a variety of convenient ways. We do not, therefore, go into more detail about aberration calculations here, but we shall give a few examples to show what may be expected in simple cases.

### 6.7.1 Spherical Aberration of Thin Lenses

A thin lens of power $K$ will bring light from an object at infinity to a focus at its second principal focus $F'$, as in figure 6.18, in the Gaussian optics approximation. The distribution of curvature between the two surfaces of the lens can be changed without altering the Gaussian properties, but this does change the aberrations. The process of redistributing curvature is called *bending* the lens, and figure 6.19 shows the effect of bending on the spherical aberration of a thin lens of positive power. The curve is scaled down vertically with increasing refractive index, and it is moved bodily vertically and horizontally with change of conjugates, i.e., if the object is not at infinity. A lens of negative power with corresponding bending would have a similar set of curves reflected about the horizontal axis.

From such curves, combined with raytracing to take account of finite glass thicknesses, arises the cemented doublet approximately corrected for Seidel spherical aberration by cancellation of positive and negative contributions; by choice of suitable dispersions of the two glasses, it may also be corrected for longitudinal color, as in figure 6.20.

### 6.7.2 Aplanatic Doublets

The doublet of figure 6.20 has two aberrations corrected by use of two degrees of freedom: bending and the ratio of dispersions of the components. To correct for coma as well, a third degree of freedom is needed. This is obtained either by careful choice of the ratio of refractive indices of the

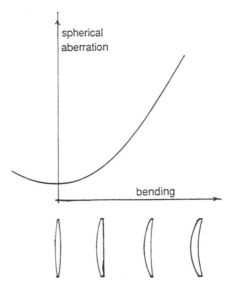

6.19 Changing the spherical aberration of a thin lens by "bending" it. The power is held constant by adding the same increment of curvature to each surface.

6.20 An achromatic doublet. The powers of the two components are chosen to cancel the dispersions, and the overall bending is chosen to cancel the spherical aberration (but there is always higher-order aberration, which is balanced against low orders to leave some residual spherical aberration).

components (but this depends on the glass types available) or by breaking the contact between the components (figure 6.21). Such a system, corrected for Seidel spherical aberration and coma, is said to be *aplanatic*.

### 6.7.3 Astigmatism and Field Curvature Correction

The doublets described in sections 6.1.7 and 6.7.2 are well corrected over only a narrow angular field, one or two degrees at most, depending on the focal length and relative aperture. Astigmatism and field curvature intervene at larger field angles. Field curvature cannot be corrected in a "thin"

6.21 An aplanatic (coma-corrected) doublet. The split contact gives an extra degree of freedom, which permits coma correction.

system since it is simply proportional to the power of the system. Likewise, astigmatism is proportional (with a different constant) to the power of a thin system if the aperture stop is in contact with the system. It is possible to correct Seidel astigmatism in a thin system by putting in heavy coma and spherical aberration and moving the aperture stop away from the lens so that the contributions due to stop shift from these aberrations cancel the astigmatism. However, such a system is of restricted use since the relative aperture has to be very small. In general, astigmatism and field curvature are controlled by using "thick" systems, i.e., components separated by air spaces. These considerations lead to a variety of systems such as photographic lenses and projection lenses of which one of the simplest and oldest types is shown in figure 6.22.

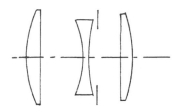

6.22 The Cooke triplet, designed by H. Dennis Taylor. This was the first system designed on the basis of aberration theory to obtain correction of astigmatism and field curvature.

### 6.7.4 Symmetrical Systems

If an optical system has a plane of symmetry as in figure 6.23, with the aperture stop at the plane of symmetry and with equal conjugates so that the magnification is −1, then all aberrations depending on odd powers of the field coordinate are identically zero. That is, coma, distortion, and transverse color of all orders, not only the Seidel terms, are zero. This result is utilized in the design of copying lenses.

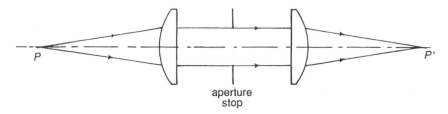

6.23 A system with symmetry about a central plane, the plane of the aperture stop. Such systems have no odd aberrations if the object and image positions are also symmetrical about this plane.

### 6.7.5 Aspheric Surfaces

So far it has been tacitly assumed that the reflecting or refracting surfaces in an optical system are spherical (or plane). The very good reason for this assumption is that such surfaces are much easier to make to the accuracy needed in optics (of the order of the wavelength of light) than surfaces of any other shape, because of the nature of the optical polishing process. However, nonspherical or *aspheric* surfaces offer some advantages in optical design. The main advantage can be illustrated by considering the Schmidt camera, an astrographic instrument. It is easily shown that a mirror in the shape of a concave paraboloid of revolution forms an image free from spherical aberration of an object at infinity on its axis, as in figure 6.24a. Such a mirror is commonly the primary or main mirror of a large astronomical telescope, but it has only a very small angular field of view because of coma. If the mirror were spherical, it would have large spherical aberration, as in figure 6.24b. Now suppose that instead of aspherizing the mirror to make it a paraboloid we put the aspherizing on a thin plate at the center of curvature of the mirror and use the aspheric plate as the aperture stop, as in figure 6.24c. The spherical aberration is corrected, and because the principal rays now meet the mirror normally, off-axis beams are treated just the same by the mirror as the axial beam and there is no coma. This is the principle of the Schmidt camera, an idea that has been extended in several ways to other systems with aspherics.

In spite of the success of the Schmidt camera and its derivatives, aspheric surfaces do not solve all optical design problems; in fact very few optical systems have more than one aspheric surface. This is partly because aspherics are very costly to make but also partly because correcting spherical aberration seems to be the main problem which aspherics can solve easily. But there are many other aberrations to be dealt with in systems of large field coverage.

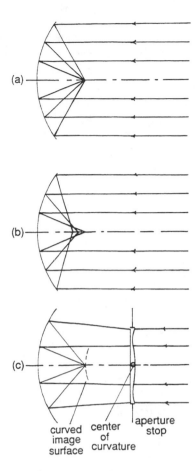

6.24 Concave mirrors as imaging elements. (a) A paraboloidal mirror has an unaberrated point image for an object at infinity on the axis. (b) A spherical mirror has large spherical aberration under the same conditions. (c) In the Schmidt camera the aberration of the spherical mirror is corrected by a thin aspheric plate at the center of curvature; this is also the stop plane, and the system then has good aberration correction over a relatively wide field.

## 6.7.6 Reflecting Systems

If an optical system consists only of mirrors, there is no chromatic aberration (at least in the geometrical optics sense; as will be seen in chap. 7, there are small physical optics effects which are a kind of chromatic aberration). Coupled with this great advantage there is the disadvantage that mirrors get in the way of each other along the optical axis so that there are very few reflecting optical systems containing more than two mirrors (i.e., convex or concave mirrors; plane mirrors for redirecting the optical axis are not relevant in this context). This considerably restricts the range of possible designs. A typical system which illustrates these points well is the Cassegrain telescope (figure 6.25). Both mirror surfaces in the telescope are aspheric, the concave being paraboloidal and the convex hyperboloidal. The imaging cones of rays are hollow (the technical term is "central obstruction"), and the angular field is restricted not only by off-axis aberrations but by the two mirrors "shearing" against each other and producing a distorted and asymmetric pupil.

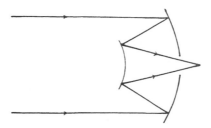

6.25 The Cassegrain telescope. The concave mirror is a paraboloid, and the convex is a hyperboloid.

In some applications the problems of central obstruction and sheared pupils can be avoided by using beam-splitting surfaces. Figure 6.26 shows how the good imaging qualities of a concave spherical mirror used with object and image at the plane of the center of curvature could be utilized. Using the beam-splitting cube means that only 25% at best of the available light is used: a situation that would be unthinkable in an astronomical context but in instruments where laser illumination is used may not be a serious disadvantage. (Aberrations introduced by the cube itself have to be taken account of in such arrangements.)

## 6.8 Optical Design

By *optical design* as opposed to optical system design we mean the design of a single component of a system, e.g., an objective, an eyepiece, or a relay lens. When a complete optical system is designed, from light source to detector, requirements arise for the individual components, that are roughly of two kinds: (a) Gaussian optics requirements, such as focal length, magnification, numerical aperture or F/number, and field angle, and (b) aberration correction requirements. To some extent these requirements can be divided between the components, but generally considerations such as compactness dictate the layout. Such components may be available commercially, but

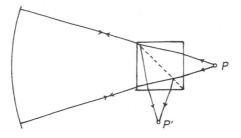

6.26 A concave spherical mirror used with object and image at its center of curvature.

if not the task is to produce optical designs for each component that fulfil these requirements as nearly as possible. There are two parts to this task. It is first necessary to decide on the general combination of optical elements that might be suitable; e.g., would a simple cemented achromatic doublet do, or perhaps something like a double-Gauss photographic objective or even something entirely new, as when Bernhard Schmidt invented the Schmidt camera? Needless to say, experience and expert advice are needed at this first stage. The second stage is more mechanical. Having decided on a general type of lens or mirror combination, it must be refined to comply with the Gaussian and aberrational requirements. Nowadays this can be done (provided the selected type of design is suitable) by means of any of several commercially available programs. These optical design programs permit a variety of constraints and tolerances to be incorporated, after which an optimization routine is carried out using a merit function which is usually a weighted sum of squares of aberrations and constraints. Given a good starting point and a realistic specification, it is thus possible to arrive at an optical design consisting of a list of optical glass types, curvatures, glass thicknesses, and air spaces which can be turned into hardware by a specialist manufacturer.

# 7

# Physical Optics and the Limits of

# Image Formation

Geometrical optics alone does not always give enough information about an optical design to predict its performance in imaging fine detail, i.e., *its resolving power* or *resolution limit*. Thus, if we we look at the spot diagram (see section 6.1) from a poor-quality design, we can get a good notion of what the image quality will be. However, the spot diagram for an aberration-free system in monochromatic light at the correct focal plane consists of a single point, suggesting infinitely fine resolution, but obviously this cannot be true, because of the wave nature of light. We therefore must examine the effect of the finite wavelength, and in this chapter we shall look at point spread functions in terms of physical optics. We use scalar wave theory, in which the field is represented by a single complex wave amplitude (usually thought of as corresponding roughly to the electric field strength in the electromagnetic wave). The theoretical details are available in many texts (e.g., Goodman 1968; Born and Wolf 1959).

## 7.1 The Aberration-free Point Spread Function

Figure 7.1 shows a system, supposed aberration-free, imaging a point $P$ at $P'$. The extent of the phasefronts in the imaging pencil is limited by the aperture stop, usually inside the system, but for convenience we replace this by its image, the exit pupil, in the image space. The wave disturbance in the region of the focus $P'$ is then regarded as due to diffraction of the phasefronts at the exit pupil. Let the diameter of the pupil be $2a$, let the distance from the pupil to the image plane be $R$, and let the wavelength of the light be $\lambda$. Then the complex amplitude in the image plane, normalized to unity at the center of the pattern, is

$$\frac{2J_1\{2\pi\alpha\eta/\lambda\}}{2\pi\alpha\eta/\lambda}, \tag{7.1}$$

where $\eta$ is the radial distance from the center of the image point, $J_1$ is the Bessel function of the first kind and first order, and $\alpha = a/R$ (small-angle approximation). This function is plotted in figure 7.2 against the scaled dimensionless parameter

$$z = 2\pi\alpha\eta/\lambda. \tag{7.2}$$

The light intensity is the squared modulus of the complex amplitude so that, again normalized to unity at the center, it is

$$I(z) = \{2J_1(z)/z\}^2. \tag{7.3}$$

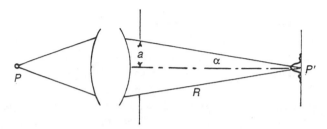

7.1 Notation for calculating the point spread function.

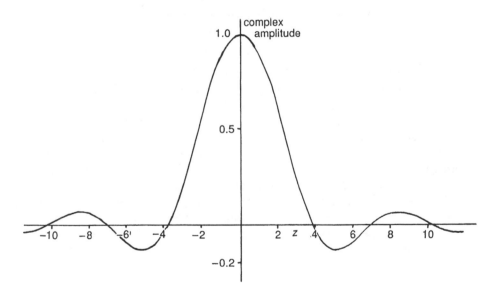

7.2 The aberration-free complex amplitude point spread function from a circular pupil with uniform amplitude. The abscissa scale is in so-called z-units, defined by eqn. (7.2).

Figure 7.3 shows equation (7.3) on a linear scale, and in figure 7.4 the ordinate is on a logarithmic scale. On account of the axial symmetry, the light intensity distribution is a series of rings around a central bright core. This is shown in figure 7.5 as a perspective plot with light intensity along the vertical direction. This light intensity distribution, known as the Airy pattern, is used as the basis of the *two-point resolution criterion*, a frequently used way of expressing the performance of an imaging system. Let the system form an image of a pair of points of equal intensity in monochromatic light, and let the two points be separate sources so that their images are incoherent with each other and cannot interfere. As the points move closer together their images begin to merge and coalesce. The distance at which they become indistinguishable is formally taken to be that at which the maximum of one falls on the first dark ring of the other, i.e., 3.83 in $z$-units. (This is a formal definition of the resolution limit. In practice it could be too coarse or too fine, depending on the working conditions and the means of measurement.) From equation (7.2) it follows that a distance in $z$-units is the same in object and image space, i.e., the magnification is scaled out, so that in real units the two-point resolution limit is in either space

$$0.61\lambda/\alpha \qquad\qquad (7.4)$$

or, if the object (or image) is at infinity, the resolution limit can be expressed as an angle:

$$1.22\lambda/2a, \qquad\qquad (7.5)$$

where $2a$ is the diameter of the exit pupil of the system. Equations (7.4) and (7.5) demonstrate that resolving power is proportional to the quantity $\alpha$ (strictly $\sin\alpha$ for large convergence angles) and inversely proportional to the wavelength; $n\sin\alpha$ is known as the *numerical aperture* when applied to systems such as microscope objectives where the space in which resolution is being considered may have a refractive index different from unity. The numerical aperture and the wavelength are therefore the main parameters governing performance in the absence of aberrations.

The fact that the Airy pattern scales directly with wavelength means that, when a system is used in polychromatic light, even if there are no aberrations in the sense of chapter 6 the point image will still show chromatic effects that are visually perceptible in well-corrected reflecting systems.

## 7.2 Image Quality with Aberrations: The Strehl Criterion

The question of how to set tolerances on aberrations, both design and constructional, depends on whether the aberrations are allowed to be large. The apparently defeatist attitude of discussing large aberrations makes sense when we consider the interaction of the detector with the optical system. Detectors such as photographic emulsions, TV camera tubes, and

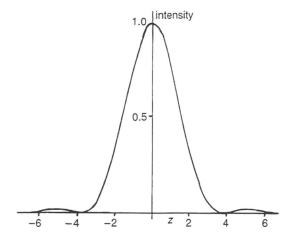

7.3 The intensity point spread function (the Airy pattern).

charge-coupled photodiode arrays have finite resolving powers of their own, and it might seem reasonable to design the optical system so that its numerical aperture corresponded to a resolution which matched that of the detector, or at any rate did not better it by too large an amount. However, resolution is not the only important parameter, and it is also desirable to have large light-gathering power. Thus it is quite reasonable for a camera lens to have a relative aperture $f/1.8$ (i.e., numerical aperture about 0.28) even though the film to be used may only resolve detail above, say, $20\,\mu$ m in scale. The high numerical aperture confers short exposure times, but the lens can have large aberrations, corresponding to a spread function approaching $20\,\mu$m, in size rather than $2\,\mu$m, as would correspond to numerical aperture (NA) 0.28 with no aberrations. We shall see in section 7.3 how tolerances are set for such cases.

Returning to small aberrations, this is the case in which, because of the properties of the detector, the full resolution corresponding to the wavelength and NA of the system can be used. Here we would ideally like zero aberrations, but design and construction limitations require that tolerances be set. The Strehl tolerance system is based on the following result: if a system starts off aberration-free (and monochromatic) and small amounts of aberration are introduced, the effects on the point spread function described in section 7.1 are approximately the same whatever the aberration type. These effects are (a) the maximum intensity at the center of the pattern decreases, (b) the zeroes of intensity in the dark rings become minima but not zero, and (c) the general shape of the central maximum remains the same and its width at half-height does not change. Of course, this does

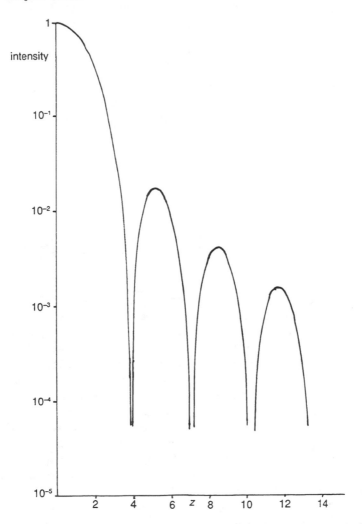

7.4 The intensity point spread function with logarithmic ordinate scale.

not hold beyond a certain point, but it suggests that a tolerance level for what is substantially a perfectly corrected system should be that for which the maximum intensity at the center does not fall below a certain level; this level is taken as 80% in the Strehl tolerance system. The amounts of different aberrations corresponding to this level vary, but for our purposes a sufficient approximation is that for any combination of aberrations it corresponds to a mean-square wavefront aberration of $\lambda^2/200$. J. J. Stamnes

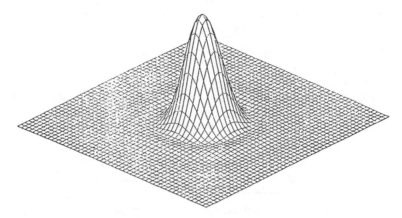

7.5 The intensity point spread function in perspective. Four squares correspond to one z-unit.

(1986) gives a detailed treatment of this topic with graphs showing many of the point spread functions. Tolerance levels for the Seidel aberrations at the Strehl tolerance limit are tabulated by many authors, e.g., Born and Wolf (1959) and Welford (1986).

Systems toleranced to the Strehl limit are sometimes described as "diffraction-limited," but in fact it is quite possible for a skilled observer or a good electronic detection system to tell the difference between such a system and one with zero aberration in a point image. Examples of systems where performance significantly better than the Strehl tolerance limit is desirable are astrometry telescopes for use in earth satellites, reduction copying lenses for microlithography, and the telescopes of theodolites. Such systems must be assembled with extreme care to achieve the design performance.

## 7.3 Tolerances for Systems with Large Aberrations

The relatively simple effects of small aberrations on the point spread function described in section 7.2 do not hold for large aberrations such as those permitted in optical systems like camera lenses. For these systems the tolerances for specifying performance are based almost universally on the *optical transfer function*. This function is defined in terms of the image produced by the system of an object with sinusoidal intensity distribution. Let the light intensity in the object be

$$I(\eta) = I_0\{1 + \cos 2\pi s\eta\}, \tag{7.6}$$

where $\eta$ is a distance coordinate in the object plane and $s$ is a *spatial frequency*; thus the object can be regarded as a grating with sinusoidal intensity distribution. Then, subject to conditions explained below, the image of this object will be of the form

$$I'(\eta) = I'_0\{1 + M \cos 2\pi s(\eta + \eta_0)\}, \tag{7.7}$$

where $|M| < 1$ and $\eta_0$ is a phase-shift term. Thus the image is similar to the object but reduced in contrast according to the factor $M$ and phase-shifted. The phase shift is usually ignored, and the image quality is then specified by the *modulation transfer function* (MTF), the factor $M$. This, of course, is a function of the spatial frequency $s$ and of the position in the field of view, the focal setting, the direction of the bars in the grating, and the wavelength spread of the illumination, but manufacturers quote MTF with these parameters specified.

The conditions under which equation (7.7) represents the image intensity distribution are (a) that the illumination in the object be incoherent (see section 7.4) and (b) that the image formation be isoplanatic. The latter is a condition on constancy of the aberrations over the part of the field of view where the MTF is to be measured or calculated. In practice it means that the form of the point spread function should not change significantly over this field. Even ignoring these conditions, the MTF gives a useful impression of performance: the reason for this is that the MTF as a function of spatial frequency is essentially a one-dimensional Fourier transform of the intensity point spread function so that it contains coded information about the point spread function. In a more advanced approach a two-dimensional Fourier transform of the intensity point spread function is taken and the variables in this transform, $s$ and $t$, say, correspond to components of spatial frequency in two orthogonal directions.

The MTF is measured either by scanning the point spread function with a slit and then taking the Fourier transform or by scanning the image of a suitable sinusoidal test chart. In either case the appropriate illumination and so forth are used. For a given optical system design the MTF can be calculated as a Fourier transform of the point spread function, and this itself is obtained by a numerical diffraction calculation for which computer programs are available. The calculation is quite lengthy, and it is sometimes quicker to use an approximation based on the spot diagram, which can be regarded as a geometrical optics point spread function; this approximation is also incorporated in optical design optimizing programs. More details about MTF calculation are given, e.g., by Gaskill (1978).

The form of the MTF for an aberration-free system with a circular pupil and in approximately monochromatic light is shown in figure 7.6, where it can be seen that the MTF falls to precisely zero at a spatial frequency $s_{\max}$ given by

$$s_{\max} = 2\alpha/\lambda. \tag{7.8}$$

As before, $\alpha$ is the convergence angle in the image space. This limit corresponds roughly to the reciprocal of the size of the point spread function. The effect of aberrations is to lower the plot shown in figure 7.6 for all points beyond $s = 0$, but the final cutoff spatial frequency is still as given by equation (7.8).

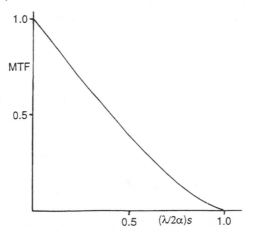

7.6 The modulation transfer function for a circular pupil with no aberrations.  The abscissa scale is normalized so that it reads unity at the cutoff spatial frequency of $2\alpha/\lambda$.

MTF is, of course, an optical analogue of the frequency response or transfer function of an electrical network.  In a network the frequency response is the Fourier transform of the impulse response function, and correspondingly the optical transfer function is the Fourier transform of the point spread function.  Viewed in this way it can be seen that the MTF can be cascaded for systems in tandem, but there is a restriction: the connection between these systems must be incoherent; that is, it must be truly the *intensity* which is transferred.  Thus consider, as a reductio ad absurdum, two aberration-free optical systems in tandem, as in figure 7.7, so that the image of system 1 is the object for system 2.  The two systems together form one aberration-free system, since the fact that there happens to be a real image between them is irrelevant.  Thus the MTF is again as in figure 7.6, and it is *not* obtained by taking the square of the ordinates in figure 7.6. However, if there were a diffusing screen at the intermediate real image, the connection would no longer be coherent and the combined MTF would be obtained as the product of the individual MTFs.  As a more realistic example, the MTF of a TV camera lens may be multiplied by that of the camera tube and again by that of the rest of the electronics, since each of these subsystems is connected incoherently with its neighbor.

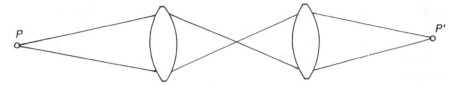

7.7 Optical systems in tandem. The MTF is *not* the product of the individual MTFs.

The case mentioned above of a TV camera tube is a good example of the use of the MTF concept. A 525-line TV system needs to resolve just that number of lines in the image and no more, so that the camera lens should be designed to have as good an MTF as possible up to the corresponding spatial frequency in the image but beyond that its performance is irrelevant. But it is still true that, other things being equal, the focal ratio ($f$-number) should be as small as possible for light-gathering power so that the aberrations will be much larger than diffraction-limited.

## 7.4 Coherent and Partially Coherent Illumination

So far we have been assuming that the objects of which images are being formed are either self-luminous (e.g., a lamp filament) or illuminated by polychromatic light incident over a large range of angles. Under these conditions each point of the object emits radiation which has a rapidly changing phase relationship with that from all the other points, and the effect of this is that the image is obtained by adding together the intensity point spread functions from all points of the object. In Fourier terms, the image is the convolution of the object intensity distribution with the intensity point spread function; this assumption lies behind the standard treatment of MTF.

There are many cases of practical importance where the assumption of incoherent illumination does not apply. An example is in microscopy of light-transmitting objects: here it is usual for the illuminating cones of light from the condenser to fill the aperture of the objective only partially, since this gives an image with better contrast than if the objective aperture were completely filled, and under this condition there is some degree of coherence between the light from neighboring points in the object. At this juncture we can explain what is meant by coherence in relation to this example. If the illumination is completely incoherent, then the image will be built up by superimposing intensity point spread functions from each object point by summation or integration of intensities. If it is not completely incoherent, then we cannot simply add intensities, and if the region where two neighboring point spread functions overlap is examined, then it can be seen that there are traces of interference effects between the beams. In fact the phenomenological quantification of the degree of partial coherence is in terms of the contrast of these interference effects. The contrast depends on the distance between the two points, and it leads to the concept of a "coherence patch," a small region within which light from any two points will show interference effects if suitably combined. In our example of microscopy the coherence patches are generally not very much larger than the resolution limit (eqn. 7.4) if ordinary thermal illumination is used, and thus the effects of the partial coherence are only perceptible if detail near the resolution limit is examined.

On the other hand in illumination with a single transverse-mode laser the illumination is completely coherent over the object and the effects of coherence are very marked indeed. The image is built up by summing or integrating the *amplitude* point spread functions from each point of the object to obtain the complex amplitude distribution in the image, and what is seen or measured is the intensity, i.e., the squared modulus of the complex amplitude. There is a spatial frequency response formalism for completely coherent illumination (see, e.g., Born and Wolf 1959), but this is not as useful in discussing image quality as the incoherent case since the former deals only with response to spatial frequencies in complex amplitude.

There is another reason why the frequency response formalism for coherent illumination must be used with care—the phenomenon usually called *laser speckle*. Figure 7.8 shows a ground-glass diffusing screen illuminated from the rear and imaged by an optical system. In incoherent illumination the image would appear to be of uniform intensity, but with laser illumination the image has the grainy speckly appearance familiar to anyone who has seen the effect of a HeNe laser illuminating a rough surface. The speckle is a random interference pattern between light scattered from the different points of the screen with random phase differences, and it acts very effectively in obscuring the detail of an image in the form of an intensity distribution. The statistics of the amplitude and intensity distributions in speckle formed under different conditions has been studied exhaustively (see, e.g., Dainty 1984). Here it is sufficient to note that, if the surface is rough enough to produce phase variations exceeding $2\pi$, the speckle will have very high contrast with many areas of apparently zero or near zero intensity. In fact the probability density distribution of light intensity is as in figure 7.9, from which it can be seen that zero *is* the most probable intensity. Clearly one should always try to avoid the need to form images of extended objects in coherent illumination unless the phase structure which contributes speckle is the object of the study.

In principle even a slight degree of coherence of illumination will show

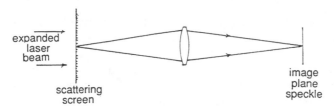

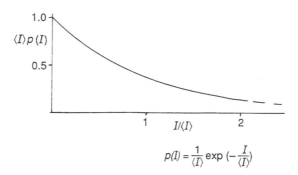

7.8 The formation of image-plane speckle.

$$p(I) = \frac{1}{\langle I \rangle} \exp \left( -\frac{I}{\langle I \rangle} \right)$$

7.9 The probability density function for the light intensity in a speckle pattern formed from a strongly scattering surface.

phase interference effects of the same fundamental nature as speckle, but these effects are unnoticeable in most practical cases of illumination by thermal sources. If, for whatever reason, it is essential to use a laser to illuminate an object in an optical system, there are methods available for reducing the coherence to the extent necessary to minimize speckle effects. Some of these are discussed in chapter 8.

## 7.5 Optical Systems and the Fourier Transform

Fourier series and Fourier transforms appear naturally in mathematical treatments of all wave phenomena, since the (mathematically) simplest wave form is sinusoidal and Fourier methods permit more complex wave forms to be synthesized by the addition of sine waves. Thus it is not surprising that there has been much mention already in this chapter of Fourier transforms. In this section we give a more general treatment of one of the aspects of optics to which Fourier theory most aptly applies.

Figure 7.10 shows an aperture in an opaque plane screen with a monochromatic plane wave incident on the screen. The screen diffracts the light, and according to simple scalar diffraction theory the complex amplitude diffracted into the direction with components $(u, v)$ is proportional to

$$\iint \exp\{(2\pi i/\lambda)(ux + vy)\}\, dx\, dy, \qquad (7.9)$$

where $x$ and $y$ are coordinates in the screen and the integration extends over the area of the aperture in the screen (see, e.g., Born and Wolf 1959). Mathematically this expression is a two-dimensional Fourier transform (ignoring trivial scaling factors) of a function which is unity over the area of the aperture of the screen and zero elsewhere. Optically we can say that the complex amplitude diffracted to infinity in the direction $(u, v)$ is proportional to the Fourier transform of the (constant) complex amplitude in the plane of the aperture.

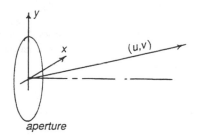

7.10 Diffraction at an aperture in a screen.

We can immediately generalize this by noting that, if the incoming wave is not plane but has a complex amplitude distribution $F(x, y)$, the diffracted complex amplitude is proportional to

$$\iint F(x, y)\exp\{(2\pi i/\lambda)(ux + vy)\}\, dx\, dy, \qquad (7.10)$$

and from this it follows that the diffracted complex amplitude is the Fourier transform of the complex amplitude in the aperture, whatever values the latter takes.

In practice it is not convenient to observe diffracted complex amplitudes and intensities at or near infinity and it is usual to circumvent this difficulty by following the screen with a lens of convenient focal length, as in figure 7.11. The diffraction pattern then appears, suitably scaled, at the focal plane of the lens.

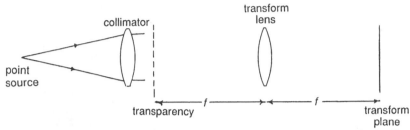

7.11 The diffraction pattern at infinity brought to a focus at a finite distance by a lens.

transform
lens

collimator

point
source

transparency ———— f ————— ⟶ ⟵ ——— f ———

transform
plane

7.12 A diffractometer.  The complex amplitude at the transform plane is the Fourier transform of that transmitted by the transparency. (The lenses are shown schematically; in practice they would be multicomponent systems.)

What is shown in figure 7.11 is, however, the same thing as a lens with an aperture forming an image of a point object at infinity so that the previous paragraph can be rephrased to say that the complex amplitude point spread function of a lens is the Fourier transform of the complex amplitude in the pupil. For example, for a lens with a circular pupil and no aberrations the expression in equation (7.1) is well known to be, with suitable scaling, the Fourier transform of a function which is unity inside a circle and zero elsewhere.  There is a small complication here: strictly speaking, the aperture in figure 7.11 should be at the first focal plane of the lens if the complex amplitude in the diffraction pattern is to be precisely the Fourier transform of that in the aperture; if this is not so, a quadratic phase term appears in the complex amplitude in the diffraction pattern, but this does not change the intensity.  This leads to systems such as that in figure 7.12 where a transparency with perhaps phase and amplitude structure is set up at the first focal plane of an aberration-free objective and the Fourier transform of the structure (as a complex amplitude distribution) is obtained at the second focal plane of the objective.  The objective is often referred to as a Fourier transform lens, although strictly speaking it is diffraction at the aperture which does the transforming and the lens is there merely to bring the transform to a convenient position for observation.  Optical systems of this kind, called diffractometers, have been used, for example, in the interpretation of X-ray crystal diffraction patterns, since these patterns are also formed by the diffraction of waves.  Also, the spatial frequency content of the original transparency can be changed by appropriate manipulations at the transform plane, an operation known as optical processing or spatial filtering. However, it must be said that except for a few special purposes computer processing of images is nowadays much more versatile than optical processing.

# 8

# Illumination for Image-forming

# Systems

## 8.1 Radiometric Concepts

Radiometry is concerned with the measurement of light as it affects the response speed of detectors, the exposure time through camera lenses, and so forth. The basic quantity is *flux*, which has the dimensions of power and is measured in either watts or photons per second; for many purposes the wavelength or frequency range must also be specified. Flux density, or *irradiance*, is power per unit area falling on a defined surface. Radiance refers to the radiation emitted by a source and is flux per steradian solid angle per unit area normal to the direction of view. These three quantities allow the handling of most questions of radiometry in instrumentation. Flux and flux density can be measured by well-defined experimental techniques. Radiance is not as easy to measure and is strictly defined only in the geometrical optics model, but it is an essential concept in dealing with radiometry for image-forming systems. By comparing the definition of radiance with the expression for the Lagrange invariant (eqn. 3.6), it can be seen that in the paraxial region of an optical system radiance is conserved along a ray, apart from reflection and absorption losses. In fact this is true along any ray, not only along paraxial rays, as will be seen in section 8.5, so that radiance is the concept to use in following the transmission of flux through a system.

### 8.1.1 The Lambertian Radiator

In modeling actual sources such as hot filaments or secondary sources such as opaque rough surfaces, it is convenient to assume that the radiance at a given point of the source is constant over all angles of view. This follows an experimental law proposed by J. H. Lambert that is *approximately* followed by many real sources and is followed exactly (in principle) by a blackbody cavity. Thus it is usual to assume that an object for an imaging system is a *Lambertian radiator*.

## 8.2 Images of Extended Objects in Incoherent Illumination

Consider a camera lens (figure 8.1) forming an image of an extended object at magnification $m$ given by

$$m = \eta'/\eta = \sin\alpha/\sin\alpha' \tag{8.1}$$

(the sines rather than the angles are used outside the paraxial region; see chap. 6). If the object is a Lambertian radiator with radiance $B$, the flux collected from an element of area $\eta^2$ is easily found to be

$$\pi B\eta^2\sin^2\alpha. \tag{8.2}$$

Neglecting reflection and absorption losses, the same total flux in image space is also, from equation (8.1),

$$\pi B\eta'^2\sin'^2\alpha. \tag{8.3}$$

Thus the irradiance at the image is proportional to $\sin^2\alpha$. Since the speed of a photographic emulsion or the response of a photoelectric detector depends on the irradiance, it is $\sin\alpha$ which determines the speed or light-gathering power of the optical system as used for extended objects. This is more usually expressed as the relative aperture, or $f$-ratio, but the essential point is the same, that it is the convergence angle in image space which matters. Clearly the irradiance in the image is proportional to the radiance of the object.

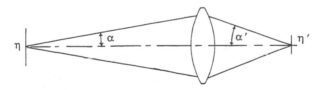

8.1 Illumination of an image of an extended object formed in incoherent illumination.

### 8.2.1 The cos⁴ Law

If a camera lens with an appreciable field of view is used with photographic emulsion, a TV tube, or any other area imaging detector, the effective exposure falls off away from the center of the field in the image. Referring to figure 8.2, the exit pupil appears elliptical from the image point, the illumination is oblique, and the solid angle is reduced.

These factors amount to a reduction in irradiance proportional to $\cos^4\beta$ where $\beta$ is the field angle. In practice vignetting of the aperture stop may make the falloff worse and a careful control of the aberrations of the imagery of the aperture stop may give an improvement, so the $\cos^4$ law is rarely followed very accurately. But it is useful as a warning, even if it is not a good approximation.

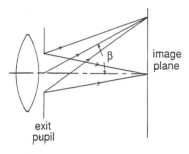

8.2 The $\cos^4$ law.

## 8.3 Condensers

Transparent or semitransparent objects such as microscope slide preparations and slides for projectors have to be deliberately illuminated for the imaging system by means of a *condenser system*. Condensers can range from a crudely molded single lens to quite elaborate multilens systems, and their design is linked to three factors: (a) the kind of light source to be used, (b) the nature of the object, and (c) the imaging system. It is convenient to consider the microscope as an example: the principles involved are easily modified and extended to cover other condenser problems. We have already noted in section 7.4 the drawbacks of using lasers as light sources for fairly conventional imaging systems (but as we shall see in chap. 13 they are very good for scanning or flying-spot systems), so it will be assumed that a thermal source such as a filament lamp or a gas discharge tube is to be used. Then from section 8.2 it follows that the radiance of the source is what controls the attainable flux density in the final image, and sources of high radiance such as quartz-halogen lamps are often used. The simplest condenser arrangement is then as in figure 8.3, where the condenser forms an image of the light source at the plane of the object. The condenser iris is usually closed down to ensure that less than the full aperture of the objective is filled with direct light, since as noted in section 7.4 this gives an image with good contrast (the rest of the objective aperture collects diffracted light from the object structure, and it is this which determines the resolving power). The obvious disadvantage of this system is that the image of the source is seen superimposed on the object. This disadvantage can be overcome by the discreet use of a ground-glass scattering screen at a suitably chosen position, but this means a loss of light flux, which may not be acceptable where physical detectors, as opposed to the eye, are used.

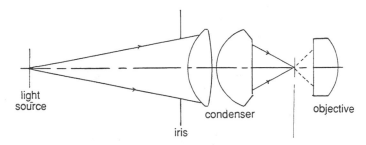

light
source

iris

condenser

objective

8.3 A microscope condenser for transmissive objects.

A version of the Köhler illumination system (figure 8.4) is used to overcome the disadvantages of the system of figure 8.3. To understand this system, we note that the standard microscope *substage condenser* has an iris diaphragm which is set at its object-side focal plane; thus the setting of this iris controls the range of illumination angles on the object and thus the degree of coherence of the illumination. The *field lens* forms an image, generally enlarged, of the source at the substage iris; thus the source image is at infinity with respect to the object on the slide. At the same time the substage condenser forms an image of the field lens iris on the slide so that the diameter of the illuminated area is controlled by the field iris. In this way the Köhler illumination system permits independent control of the degree of coherence and the size of the illuminated field, and there is uniform illumination over this field. This system can be adapted in many different ways to suit the requirements of a wide range of illumination problems. It is particularly useful where high throughput of radiation is needed, e.g., in microphotography or in projection, since there is no diffuser.

A system intermediate in complexity between those of figures 8.3 and 8.4 is that generally used for slide projectors, as in figure 8.5. In this illumination system the degree of coherence and the size of the illuminated field are both fixed but the object (the slide) is uniformly illuminated since the lamp is imaged into the entrance pupil of the projection lens.

For a source of given radiance and size the remaining variable which controls the illumination is the collecting angle of the condenser system. Thus in figure 8.5 the first component is an aspheric lens, molded in glass, since

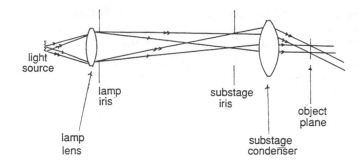

8.4 The Köhler illumination system. The illuminating NA and the area of illumination are under separate control by the two iris diaphragms.

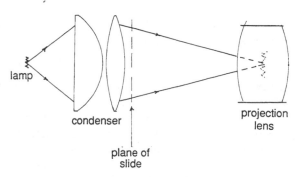

8.5 Condenser system for a slide projector. The source image underfills the pupil of the projection lens so as to make the illumination partially coherent; this produces a crisper image than would be obtained if the pupil were completely filled.

a lens with spherical surfaces in this position could not have as high a collection angle. This line of reasoning has led to the increasing use of plastic molded Fresnel lenses as condensers, particularly where large apertures are needed as in the popular overhead projector. A Fresnel lens easily beats a glass aspheric in relative aperture: the typical overhead projector Fresnel condenser works at about $f/0.3$, measured across the diagonal.

## 8.4 Lasers as Sources for Image-forming Systems

We mentioned in section 7.4 that the main difficulty in using lasers as sources for imaging systems is the formation of speckle from a surface which is diffusing but of which the image would be, in incoherent illumination, of uniform intensity. If, as can happen, it is essential to use a laser as a light source, some method of reducing the contrast of the speckle should be used if this is permissible. Such a method usually amounts to getting a suitable number of exposures (if the detector is, say, photographic emulsion), between each of which the geometry of illumination is varied in such a way as to make the successive speckle patterns independent. Then the intensities in the successive exposures are summed and the total speckle contrast is reduced. As a rough guide, if $N$ independent exposures are given, the speckle contrast is reduced by the factor $N^{1/2}$. For a detector other than photographic emulsion a method of summing or averaging the independent exposures must be found. For a closed circuit TV system this would involve a frame store, a device which can store digitally the signal strength at all 250,000 or so picture points in a TV frame.

Figure 8.6 shows the general principle of a system for reducing speckle contrast. Assuming a condenser system similar to that in figure 8.5, the illumination would produce a single bright point in the pupil of the objective (representing the beam waist of the laser; see chap. 9) surrounded by light diffracted by the detail in the object. Then a suitable system of moving mirrors or prisms is arranged to shift this bright point around inside the pupil between successive exposures. The effect is then substantially that of an extended incoherent or thermal source of extent corresponding to the region mapped out by the moving point of light.

A second, but less fundamental, problem with laser illumination is that of multiple reflections between refracting surfaces which give rise to so-called ghost images. Of course, this happens with all light sources, and antireflection coatings (see chap. 10) are used to mimimize the effects, but if the coherence length of the laser light is long enough to permit interference between the direct beam and a twice reflected beam, the resulting ghost image is covered with an interference pattern of surprisingly high contrast. Thus, if the residual intensity reflection factor after antireflection coating is $R$ per surface and if a twice-reflected beam interferes with the original beam without any change in beam diameter, the resulting fringe contrast is approximately $2R$. That is, the fringe intensity would take the form

$$I = 1 + 2R\cos\eta, \tag{8.4}$$

where $\eta$ is a coordinate across the fringe system.

The effect is worse if the twice-reflected beam is reduced in diameter, as can often happen.

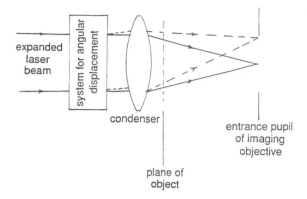

8.6 Laser illumination for an imaging system. The system for angular displacement moves the laser focus around in the pupil during a period shorter than the time constant of the detection system, thus reducing speckle effects.

## 8.5 Radiance in Geometrical Optics

The concept of radiance is useful in a more general context than that of image-forming systems. As noted in section 8.1, radiance is conserved along any ray in an optical system within the geometrical optics model, excluding losses from reflection, absorption, or scattering. This is expressed by a very general theorem which is in no way restricted to image-forming systems. Figure 8.7 shows the input and output of any optical system, with no symmetry or imaging restrictions and subject only to the condition that it is possible to trace from one input ray one or a finite number of output rays; i.e., there is no general scatter inside the system. Such a ray is indicated starting from $P_1$ in the input space and passing through $P_2$ in the output space. Coordinate systems $xyz$ and $x'y'z'$ are set up at $P$ and $P'$, and the ray directions are indicated by the respective direction cosines ($L$, $M$, $N$ and $L'$, $M'$, $N'$) with respect to these coordinate axes. The two sets of axes need not be parallel to each other, and the only restriction is that the rays not lie along the $z$ axes. Let the incoming ray be shifted in position and direction by small amounts $dx$, $dy$, $dL$, $dM$; then the following result holds for the corresponding changes to the emergent ray (see, e.g., Welford 1986):

$$n'^2 \, dx' \, dy' \, dL' \, dM' = n^2 \, dx \, dy \, dL \, dM. \qquad (8.5)$$

This result is analogous to Liouville's theorem in mechanics. It expresses radiance conservation along any ray. Integrated over areas and angles, it leads to upper bounds on the extent to which light flux can be concentrated. Equation (8.5) is the most general expression of the well-known rule that the brightness of an image can never be increased by passing it through an

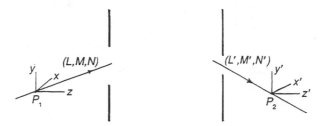

8.7 Coordinates for eqn. (8.5) expressing the generalized Lagrange invariant.

optical system, apart from a trivial factor of the square of the refractive index.

# 9

# Laser Beams

The simplest form of laser beam, as in the ubiquitous HeNe laser at 632.8 nm wavelength, has certain properties in addition to coherence and monochromaticity which have to be taken account of in some applications. The well-known Gaussian intensity profile persists if it is taken through a sequence of lenses along the axis, and at certain points that can be more or less predicted by paraxial optics a "focus" is formed. But when, as often happens, the convergence angle in the space in which this occurs is small, say, 0.001 or less, some departures from the predictions of the paraxial optics of chapter 3 occur. In this chapter we shall examine these effects, since they are of importance in many systems, including the wide class of scanning optical systems (see chap. 13).

## 9.1 Gaussian Beams

In paraxial approximation the simplest form of a single-mode beam is the $\text{TEM}_{00}$ Gaussian beam shown in figure 9.1. Starting from the narrowest part, known as the waist, the beam diverges with spherical phasefronts. The complex amplitude at the waist has the form

$$A = A_0 \exp(-r^2/\omega_0^2), \tag{9.1}$$

where $\omega_0$ is called the beam width and $r$ is a radial coordinate.

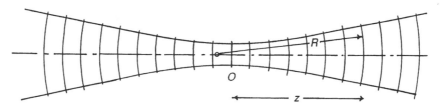

9.1 A Gaussian beam. The beam waist is at $O$. The circular arcs indicate phasefronts, and the hyperbolic curves show the radial distance at which the intensity is $1/e^2$ that of the intensity on the beam axis.

At a distance $z$ along the beam in either direction, the complex amplitude is, apart from a phase factor,

$$A = (\omega_0/\omega)A_0 \exp(-r^2/\omega^2), \qquad (9.2)$$

where $\omega$ is given by

$$\omega(z) = \omega_0\{1 + (\lambda z/\pi\omega_0^2)^2\}^{1/2}. \qquad (9.3)$$

At a distance $z$ from the waist, the phasefronts have a radius of curvature $R$ given by

$$R(z) = z\{1 + (\pi\omega_0^2/\lambda z)^2\}. \qquad (9.4)$$

The beam contour of constant intensity $(1/e^2)A_0^2$ is a hyperboloidal surface of (small) asymptotic angle $\theta$ given by

$$\theta = \lambda/\pi\omega_0. \qquad (9.5)$$

It can be seen that the centers of curvature of the phasefronts are not at the beam waist, in fact the phasefront is plane at that point. It was noted in chapter 2 that geometrical wavefronts are not exactly the same as true phasefronts, and if in this case we postulate that geometrical wavefronts should have their centers of curvature at the beam waist, we have an example of this. However, the difference is small unless the convergence angle $\theta$ is very small, or, more precisely, when the Fresnel number of the beam is not much larger than unity:

$$\text{Fresnel number} \equiv \omega/\lambda R \leq 1. \qquad (9.6)$$

There is nothing special about Gaussian beams to cause this discrepancy between phasefronts and geometrical wavefronts: a similar phenomenon occurs with beams which are sharply truncated by the pupil ("hard-edged" beams). But it happens that it is less usual to be concerned with the region near the focus of a hard-edged beam of small Fresnel number, whereas Gaussian beams are frequently used in this way. Thus Born and Wolf (1959) show in their figures 8.45 and 8.46 that the phasefront at the focus of a hard-edged beam is also plane, but with rapid changes of intensity and phase jumps of $\pi$ across the zeros of intensity.

## 9.2 Tracing Gaussian Beams

If the beam is in a space of large convergence angle, say, greater than 0.01, it can be traced by ordinary paraxial optics as in chapter 3, i.e., using the assumption that for all practical purposes the phasefronts are the same as geometrical wavefronts. In a space of small convergence angle it is necessary to propagate the beam between refracting surfaces by means of the proper Gaussian beam formulas and then use paraxial optics to refract (or reflect) the phasefront through each surface in turn. To do this, we need two more formulas to give the position, $z$, and size, $\omega_0$, of the beam waist starting from the beam size and phasefront curvature at an arbitrary position on the axis, i.e., given $\omega$ and $R$. These are

$$z = R\{1 + (\lambda R/\pi\omega^2)\}^{-1} \tag{9.7}$$

and

$$\omega_0 = \omega\{1 + (\pi\omega^2/\lambda R)^2\}^{-1/2}. \tag{9.8}$$

Equations (9.3) to (9.8) enable a Gaussian beam to be traced through a sequence of refracting surfaces as an iterative process. Thus, starting from a beam waist of given size $\omega_0$ (and angle given by eqn. 9.5), we move a distance $z$ to the first refracting surface. At this surface the beam size $\omega$ is given by equation (9.3) and the radius of curvature $R$ of the phasefront is given by equation (9.4). The radius of curvature $R'$ of the refracted phasefront is obtained by paraxial optics using the equations of figure 3.4 and taking $R$ and $R'$ as the conjugate distances $l$ and $l'$. Then the position and size of the new beam waist are found from equations (9.7) and (9.8). These procedures can be carried through all the refracting surfaces of the optical system.

It can be seen from equation (9.8) that $z$ and $R$ are substantially equal when $\lambda R/\pi\omega^2$ is very small, in agreement with the statement in section 9.1 about equation (9.6). When this is so, there is no need to use these special equations for transferring between surfaces; the iterative equations in figure 3.6 can be used, with the understanding that the paraxial convergence angle $u$ is the equivalent of the asymptotic angle $\theta$ in equation (9.5).

There are no simple equations for hard-edged beams corresponding to equations (9.3) to (9.8) for use with very small convergence angles. Numerical calculations of the beam patterns near focus have been published for some special cases, and these show, as might be expected, very complex structures near the "focal" region, however that is defined.

## 9.3 Truncation of Gaussian Beams

The theoretical origin of the Gaussian beam is as a paraxial solution of the Helmholtz equation, i.e., a solution concentrated near one straight line, the axis (see, e.g., Kogelnik and Li 1966), but although most of the power is within the region near the axis, the solution is nonzero, although very small, at an infinite distance from the axis. Thus the Gaussian profile is truncated when it passes through any aperture of finite diameter—e.g., a lens mount, an aperture stop, or even the finite-diameter end mirror of a laser resonator—after which it is no longer Gaussian and the equations of sections 9.1 and 9.2 are no longer valid! In practice this need not be a problem, for if the radius of the aperture is $2\omega$, the complex amplitude is down to 1.8% of its value at the center and the intensity is 0.03% of its value at the center. Thus it is often assumed that an aperture of radius $2\omega$ has no significant effect on the Gaussian beam, and this assumption is adequate for many purposes, although not all.

Sometimes it is useful to truncate a Gaussian beam deliberately, i.e., turn it into a hard-edged beam, by using an aperture of radius less than, say, $\omega$. In this way an approximation to the Airy pattern (section 7.1) is produced at the focus instead of a Gaussian profile waist, and this pattern may be better for certain purposes, e.g., for printers where the spot must be as small as possible for an optical system of given numerical aperture.

## 9.4 Gaussian Beams and Aberrations

In principle a Gaussian beam is a paraxial beam, from the nature of the approximations made in solving the Helmholtz equation, as explained in section 9.3. However, Gaussian beams can be expanded to large diameters simply by letting them propagate a large distance, and they can acquire aberrations by passing through an aberrating lens or mirror system. The beam is then no longer Gaussian, of course, in the strict sense, but we stress that conventional optical design ideas involving balancing and reduction of aberrations can be applied to systems in which Gaussian beams are to propagate. For example, a *beam expander*, of which one form is shown in figure 9.2, is an afocal system intended to do what its name implies: if it has aberrations as an afocal system, the output beam from a Gaussian input beam will not have truly plane or spherical phasefronts.

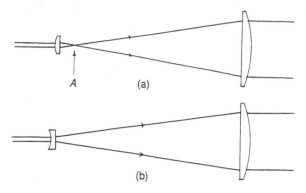

9.2 Laser beam expanders (a) suitable for low-power beams (a spatial filter to remove clutter can be placed at the focus or beam waist at $A$), (b) suitable for high-power beams, where a waist or real focus would cause air breakdown. The small lens in (a) is often a low- or medium-power microscope objective.

## 9.5 Non-Gaussian Beams from Lasers

Not all lasers produce Gaussian beams, even ignoring the inevitable truncation effects of resonator mirrors. Some gas lasers (e.g., helium-neon at any of its lasing wavelengths) produce Gaussian beams when they are in appropriate adjustment, but they can produce off-axis modes with more structure than a Gaussian beam. Other gas lasers (e.g., copper vapor lasers) produce beams with a great many transverse modes covering an angular range of a few milliradians in an output beam perhaps 20 mm across. Some solid state lasers, e.g., ruby, may produce a very non-Gaussian beam because of optical inhomogeneities in the ruby. Laser diodes, which are useful as very compact coherent sources either cw or pulsed, produce a single strongly divergent transverse mode which is wider across one direction than the other. This mode can be converted into a circular section of approximately Gaussian profile by means of a prism system, as in figure 4.11.

# 10

# Thin-Film Multilayers

Antireflection coatings, universally used on, for instance, camera lenses since the 1940s, are examples of the wide range of optical applications of thin films. Other examples are high reflecting coatings, narrow-band and broad-band filters, polarizing filters, and beam-splitters. These coatings are available in the visible, infrared, and, to a lesser extent, ultraviolet regions of the spectrum. In this chapter we explain the principles of these coatings and give examples of available coating types.

## 10.1 The Single-Layer Antireflection Coating

We first recall the expressions for the reflectance of interfaces between transparent media, the Fresnel formulas (see, e.g., Born and Wolf 1959). For incidence from medium 1 with index $n_1$ against medium 0 with index $n_0$ and with angles of incidence respectively $\theta_1$ and $\theta_0$, as in figure 10.1, the complex amplitude reflectance for $p$-polarized light (the electric field vector in the plane of incidence) is

$$r_p = \frac{n_1 \cos\phi_0 - n_0 \cos\phi_1}{n_1 \cos\phi_0 + n_0 \cos\phi_1}. \tag{10.1}$$

And for the intensity reflectance

$$R_p = |r_p|^2. \tag{10.2}$$

For $s$-polarized light (electric field vector perpendicular to the plane of incidence) the expressions are

$$r_s = \frac{n_1 \cos\phi_1 - n_0 \cos\phi_0}{n_1 \cos\phi_1 + n_0 \cos\phi_0}, \tag{10.3}$$

$$R_s = |rs|^2. \tag{10.4}$$

Putting $\phi_1$ equal to zero and $n_1$ equal to one, we obtain the well-known result that for glass of index 1.5 the intensity reflection loss at the interface is about 4%. Also, from equation (10.1) and from Snell's law it follows that, when $\tan\phi_1 = n_0/n_1$ (the Brewster angle), the reflectance for $p$-polarized light is zero.

71

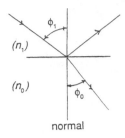

10.1 The Fresnel formulas for reflection and refraction.

Now referring to figure 10.2, a transparent substrate of refractive index $n_0$ has deposited on it a thin layer of index $n_1$ (with $n_1 < n_0$) of thickness $d$. Let light of wavelength $\lambda$ be incident normally on the layer. The beam reflected from the interface will be $2n_1 d/\lambda$ wavelengths behind that reflected from the top face, and if this retardation is half a wavelength, the two beams will interfere destructively and the reflected intensity will be less than it would have been from the uncoated substrate. If $n_1$ is chosen appropriately (as $\sqrt{n_0}$), the amplitudes of the two reflected beams will be equal and there will be zero reflection from the coated surface.

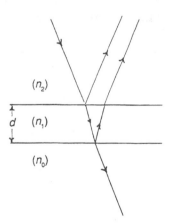

10.2 Interference in a thin film, the principle of antireflection coatings.

The above is a simple picture of the effect of a single coating, but it does not take into account multiple reflections in the layer and it does not explain how the reduced reflectivity due to interference between two back-reflected beams is accompanied by an increase in transmission; both these points will be covered in the full treatment of multilayers in section 10.2.1.

The general properties of multilayers are broadly similar to those of a single layer, so here we give a qualitative description of a few general properties of single layers.

## 10.1.1 Wavelength Dependence

If the refractive index of the layer is chosen to give zero reflectance at, say, $\lambda_0$, this will not be so at other wavelengths. At $\lambda_0/2$ the retardation will be a whole wavelength (neglecting dispersion effects in the material of the layer), and so the reflectance will be the same as for the bare substrate. Thus if, as is customary, the reflectance is plotted against the scaled reciprocal of the wavelength, the plot will look as in figure 10.3. This illustrates a general rule that films which are an integral number of half-wavelengths thick have no effect at that wavelength, keeping the same angle of incidence (see section 10.1.2 below).

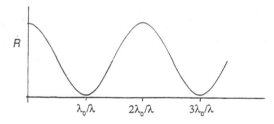

10.3 Reflectance of a single-layer antireflection coating as a function of reciprocal wavelength. The graph could equally be used to show the effect of varying the thickness of the layer, taking the abscissa as a linear scale of thickness.

## 10.1.2 Angle of Incidence

If the light is not normally incident, the retardation of the beam reflected at the interface is reduced; it is, in fact,

$$2n_1 d \cos \phi/\lambda \tag{10.5}$$

in wavelengths, where $\phi$ is the angle of incidence *inside* the film, calculated in the usual way according to Snell's law. This reduction has the effect of shifting the reflectance curve generally toward shorter wavelengths. Also, from equations (10.1) and (10.3), the amplitude reflectance at both interfaces will change with the angle of incidence and the change will be different for the two polarizations. Thus, in addition to the shift to shorter wavelengths, at oblique incidence we must consider the two polarizations separately and at large angles of incidence there will be marked differences between them.

## 10.2 General Theory

Given a multilayer with media of known thicknesses and refractive indices, the reflectance and transmittance can be calculated as a function of wavelength for both polarizations and all angles of incidence. The effect of absorption can be taken account of by assigning complex refractive indices to the materials. Figure 10.4 shows $k$ layers numbered from the substrate upward. The procedure is to assume an emergent plane electromagnetic wave of a chosen wavelength and angle of incidence. Waves in each layer in directions determined by Snell's law are assumed, and the boundary conditions for the electric and magnetic fields at each interface are written down, giving a set of $2k$ linear simultaneous equations in the $2k + 1$ unknowns. The intermediate fields inside the layers can be eliminated, and the ratio of the incident and reflected fields is obtained from a product of $2 \times 2$ matrices, one for each film. The details are given by, e.g., Macleod (1986), and, as with raytracing, software packages are available for numerical computation. The equations are summarized in section 10.2.1 below. However, as with the design of image-forming systems, the inverse problem, to design a multilayer of given properties, is not straightforward and in most cases design proceeds by numerical work starting from a few basic principles. Some of these principles are mentioned below. Again, software optimizing packages are available.

At a given wavelength and angle of incidence the addition of an integral number of half-wavelengths of optical thickness to any nonabsorbing layer does not change the reflectance or transmittance. (But it *does* change these properties at other wavelengths and/or angles of incidence.)

If the multilayer contains an absorbing component, the transmittance is the same from either direction but the reflectance will in general be different. For example, a thin metal film on glass has a lower reflectance from the glass side than from the air side.

On going from normal incidence to oblique, the first effect is a general shift of reflectance and transmittance curves to shorter wavelengths.

The Fresnel formulas (eqns. 10.1 and 10.3) show that at normal incidence there is a phase change of $\pi$ on reflection when the light is incident from a medium of low refractive index on a medium of higher index, but there is no phase change when the incidence is from the higher-index medium (assuming no absorption in either medium). This is also true at oblique incidence, except that for the $p$-polarization there is a phase change of $\pi$ on passing through the zero of reflectance at the Brewster angle.

The periodicity of behavior with respect to film thickness indicates that there is in practice a limit to the range of wavelengths over which it is possible to specify the reflectance for a proposed design, even using incom-

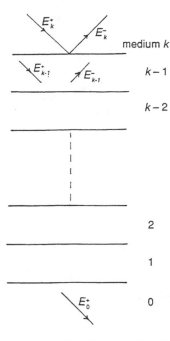

$E_k^+$   $E_k^-$

medium $k$

$E_{k-1}^+$   $E_{k-1}^-$

$k-1$

$k-2$

2

1

$E_0^+$    0

10.4 Notation for computing the properties of a multilayer.

mensurable thicknesses of layers. This range is usually regarded as given by a factor 2, an octave of frequencies.

### 10.2.1 Computing the Properties of a Given Multilayer

The layers, the substrate, and the upper medium are given subscript numbers as in figure 10.4. In each medium the refractive index is $n$ with an appropriate subscript, the angle of incidence is $\phi$, the tangential component of the total electric field (for both directions of travel) is $E$, and that of the magnetic field is $H$. We define two generalized refractive indices for the two polarizations:

$$u_p = n/\cos\phi, \qquad u_s = n\cos\phi, \qquad (10.6)$$

and we define the phase thickness of each layer

$$g = (2\pi/\lambda)nd\cos\phi, \qquad (10.7)$$

where $d$ is the metrical thickness of the layer.

Then each layer is represented by its matrix

$$A_j = \begin{pmatrix} \cos\phi & i(1/u)\sin\phi \\ iu\sin\phi & \cos\phi \end{pmatrix}, \qquad (10.8)$$

and the complete multilayer is represented by the matrix product

$$A = A_k A_{k-1} \cdots A_1. \tag{10.9}$$

The total fields in medium $k + 1$ are given by the column matrix

$$\begin{pmatrix} E_{n+1} \\ H_{n+1} \end{pmatrix} = \mathbf{A} \begin{pmatrix} 1 \\ n \end{pmatrix}, \tag{10.10}$$

assuming that the $E$ and $H$ fields in medium zero are, respectively, 1 and $n_0$.

Finally, the incident and reflected electric fields in medium $k + 1$ are given by

$$2E_{n+1}^+ = E_{n+1} + (1/u_{n+1})H_{n+1},$$
$$2E_{n+1}^- = E_{n+1} - (1/u_{n+1})H_{n+1}, \tag{10.11}$$

and the intensity reflectance of the multilayer is

$$R = |E_{n+1}^-/E_{n+1}^+|^2. \tag{10.12}$$

The two polarizations are computed separately at oblique incidence using the appropriate values of the generalized refractive indices given by equation (10.6).

It can be seen that in the above process we obtain the *steady-state* solution for the propagation of electromagnetic waves in the multilayer. This is the explanation of the paradox mentioned in section 10.1, that the transmission is increased by an antireflection coating: a solution for transients (very short pulses of light) propagated through a very thick layer would show that initially there is not increased transmission.

All of the above formalism strictly applies only to infinitely extended plane waves and optical components, but cases where this proviso matters are rare in practice.

The software packages used for these calculations may also give other useful information, such as the distribution of electric field strengths in the standing waves inside the components of the multilayer.

## 10.3 Types of Multilayer

In this section we indicate briefly the properties obtainable with different types of multilayers. These properties depend greatly on the available refractive indices and dispersions of film materials (although dispersion is not often taken into account and the indices of films as produced are rarely known to better than the second decimal place). Macleod (1986) lists many of the materials available, with indications of the method of deposition and the wavelength range of transmission.

## 10.3.1 Antireflection Coatings

The lowest index available (apart from certain coatings which are made with submicrometer voids to give an effective lower index than the bulk material) is about 1.35, which indicates that a single layer can only produce zero reflectance on a substrate of index higher than about 1.8 (figure 10.5). However, the reflectance curve is fairly broad, and for this reason (among others) optical designs are sometimes carried out entirely in high-index glass so that the coating can be simple and robust.

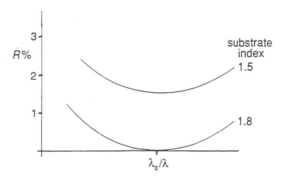

10.5 Single-layer antireflection coatings.

By using two layers, one of lower index than the substrate and the other of higher, it is possible to design for precisely zero reflectance at a given wavelength (although the coatings available are more likely to have a minimum reflectance of at least 0.001). Away from the minimum the reflectance increases more rapidly than for a single layer, as in figure 10.6, and such coatings are sometimes called V-coats for this reason; they are particularly good for components used in laser light.

The so-called broad-band antireflection coatings have three or more layers, and they can have two minima and low reflectance over, e.g., most of the visible spectrum, as in figure 10.7. However, the greater the number of layers, the more rapidly the properties change with angle of incidence; this is true for most multilayer devices.

## 10.3.2 High-Reflectance Multilayers

The principle of the high-reflecting multilayers used in laser resonator mirrors and for other purposes is illustrated in figure 10.8. A stack of layers of alternately high and low refractive indices, ($n_h$ and $n_l$), perhaps 11 or more altogether, is made with each film $\lambda/4$ in optical thickness at some chosen design wavelength. Then, recalling the phase changes mentioned in

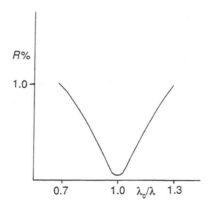

10.6 *V*-coating suitable for laser systems. The coating would contain two layers.

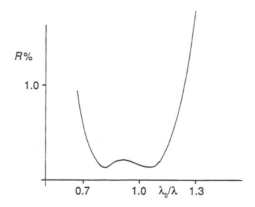

10.7 Broad-band antireflection coating containing three or more layers.

section 10.2, it can be seen that the beams reflected from each interface are all in phase and so the reflected intensity builds up as more layers are added. Figure 10.9 shows the reflectivity of such an 11-layer structure at normal incidence, computed according to the analysis of section 10.2.1. The maximum reflectance approaches indefinitely close to 1.0 as the number of layers increases, but the width of the high-reflectance band (the so-called stop band) depends only on the ratio of the high and low refractive indices in the layers. This width is given in terms of the scaled inverse wavelength by

$$2\Delta\lambda_0/\lambda = (4/\pi)\sin^{-1}\{(n_h - n_l)/(n_h + n_l)\}. \tag{10.13}$$

Figure 10.10 shows the same 11-layer multilayer computed for an angle of incidence of 30°. The general shift to shorter wavelengths and the higher reflectivity for s-polarization than for p-polarization are characteristic properties of such multilayers.

equiphase

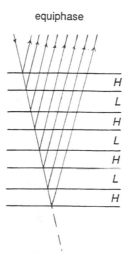

10.8 A periodic multilayer containing alternately high and low index quarter-wavelength layers. Because of the phase change effect implied by eqn. (10.1), each interface returns a reflected beam that is in phase with or $2\pi$ out of phase with the previous beam, thus building up high reflectivity.

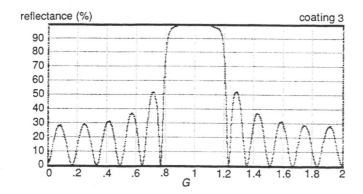

10.9 Reflectivity of a periodic multilayer with six high-index and five low-index layers, respectively, 2.4 and 1.38. The abscissa $G$ is the reciprocal wavelength $\lambda_0/\lambda$.

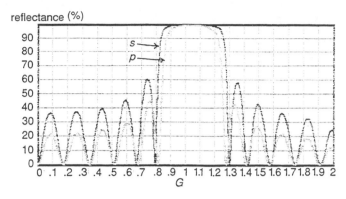

10.10 Reflectivity of the same multilayer as in figure 10.9 but at 30° angle of incidence.

The extreme indices available in the visible region are about 1.38 and 2.4, depending on conditions of deposition of the films. This restriction on width of the high-reflectance range is overcome by designs which have been computer-optimized to have different layer thicknesses (see, e.g., Macleod 1986). These designs have phase changes on reflection which vary rapidly and in a complicated way with wavelength, and this must be allowed for in any context where phase changes matter, e.g., Fabry-Perot mirror coatings.

### 10.3.3 Interference Filters

An interference filter is in effect a Fabry-Perot interferometer with spacing of at most a few wavelengths and constructed as one assembly with a thin-film layer as spacer and, usually, multilayer high-reflecting stacks as described in the previous section for mirrors. The filter is said to be of $N$th order if the spacer is $N \lambda/2$ thick. The transmission bandwidth can be as small as 0.1% of the peak wavelength, depending on the reflectance of the high-reflecting stacks. This follows from the classical Fabry-Perot theory, and the range over which there is low transmission depends on the values of refractive index in the stacks, as explained in section 10.3.2. Since the materials are all supposed to be nonabsorbing, the peak transmission of such a filter should be unity but in practice losses through scattering and absorption limit peak transmissions to about 0.8.

As with all multilayer devices, the transmission band of an interference filter shifts to shorter wavelengths and splits into two opposite polarizations as it is tilted away from normal incidence. Manufacturers can generally supply details of this effect on request; the magnitude increases with the order of the filter.

### 10.3.4 Polarizing Beam-Splitters

We know from section 10.1 that there is a Brewster angle for an interface between any two transparent media of differing refractive index. Thus, if a high-reflecting stack is built inside a prism as in figure 10.11 and set at the Brewster angle, there will be zero reflectivity for the $p$-polarization and the reflected beam will be purely $s$-polarized. Also, if there are enough layers in the stack, almost no $s$-polarized light will be transmitted and the device will act as a polarizing beam-splitter. With the materials currently available it is possible to get efficient splitting between the polarizations with a $90°$ angle between the beams. This allows a very convenient geometry in many applications.

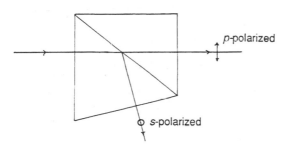

10.11 A polarizing beam-splitter. The angle is chosen to be the Brewster angle for the interface between successive layers. For a high enough refractive index of the prism pieces this angle can be $45°$.

We have mentioned only a few of the many optical applications of thin-film multilayers. For a detailed review of the types available with extensive numerical and graphical data see Dobrowolski (1978).

# 11

# Interference and Interferometry

The general topic of interferometry includes a vast field of applications and different arrangements of equipment (see, for example, Steel 1983), but in fact only a few basic principles are involved. We shall cover these by taking as an example one interferometer, that invented by A. A. Michelson, and analyzing its performance in various modes.

## 11.1 The Michelson Interferometer

Figure 11.1 shows the Michelson interferometer as it would be set up today to give the circular fringes Michelson described. The two mirrors are adjusted so that the image of one in the beam-splitter is parallel to the other and they are a distance $z$ apart. The extended incoherent monochromatic source (e.g., a filtered mercury lamp) is, in effect, at infinity with respect to the interferometer, and the objectve lens forms fringes at its focal plane which, again, are at infinity with respect to the interferometer and are therefore in the plane of the image of the source. The fringes map the function

$$2(z/\lambda)\cos\theta, \tag{11.1}$$

where $\theta$ is the angle at which a collimated beam from a point on the source meets either mirror; that is, points in the source image for which this function is an integer are bright, and points at which it is an integer plus 1/2 are dark. If $z$ is increased by moving either mirror, all the fringes expand and new fringes appear from the center; the opposite happens if $z$ is decreased. Because the fringe parameter for given $z$ is $\theta$, these fringes are called *fringes of equal inclination*, or *Haidinger fringes*.

In the alternative mode of use (figure 11.2) a small source is used (e.g., by closing down an iris in front of the extended source), the mirrors are brought approximately together ($z \approx 0$), and a slight inclination is given to one of them so that there is in effect an air wedge between the two mirrors. Then, if the eye is placed at the focus of the objective, this objective becomes a field lens and straight wedge fringes are seen at the mirrors. These fringes map the same function (eqn. 11.1), but $\theta$ is now zero and the fringe parameter is $z$. These are now *fringes of constant optical path*, or *Fizeau fringes*.

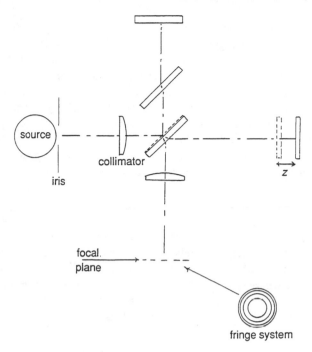

11.1 The Michelson interferometer arranged to give circular fringes at infinity with re-spect to the interferometer space.

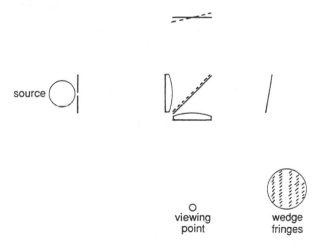

11.2 The Michelson interferometer arranged to give wedge or tilt fringes in the region of the mirrors. Each fringe corresponds to half a wavelength change in thickness of the air wedge.

There is a reciproal relationship between the source region (and regions conjugate to it) and the region of the mirrors for fringes in either position to be seen clearly. If the tilt between the mirrors exceeds about one fringe over the mirror width, then the Haidinger fringes lose contrast, since at any given point in the Haidinger fringe system the radiation arrives with a range of path differences exceeding a wavelength. Simlarly, if the source iris is opened up enough to show more than one Haidinger fringe, the Fizeau fringes lose contrast, since the light intensity at a given point in the Fizeau fringe system is the sum of that in fringes formed with path differences exceeding a wavelength. All two-beam interferometers used with extended incoherent sources display a duality of this kind between two regions, which may be called the source region and the object region.

Another factor affecting fringe contrast is the coherence length of the source. This is not a precisely defined quantity, but for the present purposes it is the optical path difference between the two arms of the interferometer at which the fringe visibility has dropped to, say, 10% when it is operating in the Haidinger fringe mode and when the mirrors are parallel enough not to affect the visibility of the Haidinger fringes. To an order of magnitude the coherence length is $\lambda^2/\Delta\lambda$, where $\Delta\lambda$ is the spectral width of the supposedly monochromatic source.

These aspects of two-beam interferometers were described systematically and in detail by Steel (1983) in a study of all aspects of interferometry.

### 11.1.1  The Michelson Interferometer as a Spectrum Analyzer

Suppose the interferometer is set up to produce Haidinger fringes and an iris at the fringe system is closed down enough to permit only the central fringe of the Haidinger system to appear. Then, as one mirror is translated a distance $z$ as indicated in figure 11.1, a detector behind the iris will record only a sinusoidal variation of light and dark for a truly monochromatic source. Now let the source be polychromatic with spectrum $G(\nu)$, where $\nu$ is the frequency of the light. Then the total signal recorded as a function of $z$ will be

$$F(z) = \int G(\nu)\{1 + \cos 4\pi\nu/c\}\,d\nu. \tag{11.2}$$

This recorded signal, known as the fringe function, is the cosine Fourier transform of the spectrum, apart from an additive constant, and thus the spectrum $G(\nu)$ can be obtained.

This technique has come to be known as Fourier transform spectroscopy. It was described by Michelson (1902), who used it, among other purposes, to select the cadmium red line as the narrowest spectrum line then available and therefore most suitable as a length standard. Michelson built a machine, the "harmonic analyzer" for finding the transform; it is described

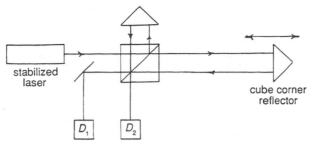

11.3 Principle of a distance-measuring interferometer. The signals from the two detectors $D_1$ and $D_2$ can be combined to show which way the cube corner is moving, provided the beam-splitter cube surface has the right phase-changing properties.

in the above reference.

## 11.1.2 Distance and Phase Measurement

With the interferometer in the Fizeau fringe mode, the shape of the fringes will indicate when one of the mirrors is not truly plane. These Fizeau fringes are no longer straight wedge fringes, and they may be regarded as forming a contour map of the surface topography with contours at $\lambda/2$ intervals. The base plane of the contours is tilted by applying a tilt to either mirror. Thus the shape of a nominally plane mirror can be checked by placing it in either arm of the interferometer.

In the same way a nominally plane-parallel window can be tested by placing it in either beam. The fringes are then contours of the function

$$2(n - 1)t, \qquad (11.3)$$

where $t$ is the thickness and $n$ is the refractive index, both possibly slowly varying functions of position in the window.

The light source for such applications as the above would be chosen to have a coherence length greater than the distances to be measured. This leads to the use of lasers for long-distance measurement in simple modifications of the Michelson interferometer. Figure 11.3 shows the principle of systems of this kind: the unexpanded beam from a frequency-stabilized laser is split and returned by cube-corner prisms (see chap. 4), and one prism is translated along the distance to be measured, which may be several meters. The cube-corner reflectors make it possible to separate the return path from the incoming path. Thus both fringe systems formed when the beams recombine may be detected. Then the signals from the fringe-counting detectors can be combined to show the direction in which the cube-corner prism is moving. The other advantage of using a cube corner instead of a plane mirror is that the fringe count is insensitve to angular movement of the prism.

### 11.1.3 The Function of the Compensator

Using a compensator, intended to verify that the composition of the two arms of the interferometer in terms of dispersive media is the same, ensures that the total optical path lengths can be equalized for all wavelengths simultaneously, and thus a perfectly black fringe can be obtained in white light for zero optical path difference. (If the coating on the beam-splitter is absorbing—i.e., metallic—there will still be some dispersion due to varying phase-shifts on reflection with wavelength.) This compensation is necessary when the interferometer is used as a spectrum analyzer, as in section 11.1.1, but is not necessary for distance measurement, as in section 11.1.2.

### 11.1.4 Testing Lenses: The Twyman-Green Interferometer

A further version of the Michelson interferometer was devised, by F. Twyman and A. Green, to measure the wavefront aberrations of lenses and other image-forming systems (Steel 1983). In this version the equivalent of Fizeau fringes are used (a small source, as in figure 11.4). One mirror is replaced by the lens to be tested and a convex spherical mirror. If the convex spherical mirror has its center of curvature at the focus of the lens and *if the lens is aberration-free*, a plane wavefront will be returned to the beam-splitter and only straight tilt fringes will be seen. If the lens has aberrations or if the convex mirror is not set with its center of curvature at the focus, the fringes will be curved and their shape will show, again as a contour map, the wavefront aberration and defocus. Many variations on this theme that have been invented for testing a variety of systems are described by Malacara (1978).

## 11.2 Interferometry and Coherence

Up to this point we have used the term "coherence" rather loosely and qualitatively. Two-beam interferometry provides a means of quantifying the definition in terms of simple experimental ideas. Figure 11.5 shows an experiment for forming two-beam fringes from an approximately monochromatic source such as a filtered mercury lamp; this is Thomas Young's famous two-pinhole interference experiment. The beams from the two pinholes spread out by diffraction and overlap with a path difference less than the coherence length of the source. Fringes with a spacing approximately equal to $\lambda/\alpha$ are formed in the region where the beams overlap. However, the contrast of the fringes depends on the size of the source, and it is found that, if

$$d > \lambda L/a, \tag{11.4}$$

where $a$ is the source size, the fringes disappear.

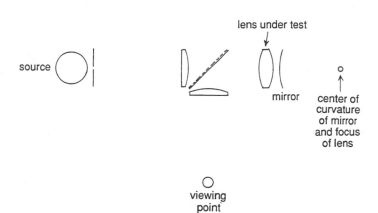

lens under test

source

mirror    center of
curvature
of mirror
and focus
of lens

viewing
point

11.4 Modification of the Michelson interferometer by Twyman and Green for measuring the wavefront aberration of lenses. The fringes map the wavefront aberration with half-wavelength contours together with any defocus or lateral shift if the center of curvature of the convex mirror and the focus of the lens do not coincide.

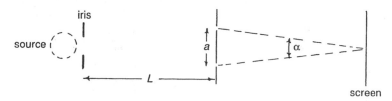

iris

source

$a$

$\alpha$

$L$

screen

11.5 Thomas Young's two-beam interference experiment interpreted as a measurement of lateral coherence.

Likewise, if the path difference between the beams is made large (exceeding the coherence length), the fringes disappear. This leads to the concept of the *degree of partial coherence* as expressed by the visibility or contrast of the fringes. The visibility is defined as

$$V = \frac{I_{max} - I_{min}}{I_{max} + I_{min}}, \tag{11.5}$$

where $I_{max}$ and $I_{min}$ are, respectively, the maximum and minimum intensities in the fringe system. Now let $I_1$ and $I_2$ be the intensity from either beam alone. Then the degree of partial coherence is

$$\gamma = \frac{I_1 + I_2}{2(I_1 I_2)^{1/2}} \cdot V, \tag{11.6}$$

so that, when the beams are of equal intensity, the degree of coherence is equal to the visibility. (This is a very simplified explanation; for a more detailed treatment see Born and Wolf 1959.) This definition takes in both of what are often called transverse coherence and longitudinal coherence, the former mainly a source size effect and the latter what we have so far called a coherence length effect. Strictly speaking, these cannot always be separated but for many practical purposes it is adequate to think in these simplified terms. Then the definition can be applied to two-beam fringes formed in any way, e.g., by sampling a beam at two different points along the same ray, to determine longitudinal coherence, as is done in effect in section 11.1.1.

We can also apply these ideas to the discussion in section 7.4 of partial coherence in image formation. Our definition of partial coherence deals with beams from two different points at which the light came originally from the same source. Thus we can consider two points in the plane of an object illuminated by a condenser system and ask about the visibility of the fringes which would be formed by light taken from the two points. As the distance between the points is increased, the coherence decreases and we could take some value, say, $\gamma = 0.1$, as defining the extent of a coherence patch (the distance around one point within which $\gamma$ is greater than 0.1). Then it is found (Born and Wolf 1959) that the size of the coherence patch decreases as the condenser numerical aperture is increased.

## 11.3 What Is Meant by Monochromatic Light?

The quantitative definition of degree of coherence introduced in section 11.2 leads to a closer examination of the notion of monochromatic light. If we wish to determine the longitudinal coherence of a beam, it is clearly ncessary that the beam should exist for a time of at least $z/c$, where $z$ is the expected coherence length (the wavetrains should be at least of length $z$). From this it follows that no light source can be perfectly coherent since this would imply at least that it had existed for an infinite time and would continue to exist, and furthermore that the complex amplitude was perfectly sinusoidal. In discussing the light from a spectrum lamp or even a helium-neon laser, we therefore speak strictly of quasimonochromatic light, implying that the complex amplitude is almost sinusoidal and changes very slowly compared with the period of the light. Thus the light amplitude at a point as a function of time could be represented by an expression of the form

$$A(t) = a(t)\cos\{\omega t - \phi(t)\}, \tag{11.7}$$

where $a(t)$ and $\phi(t)$ are random functions of time which vary slowly compared to the mean circular frequency $\omega$.

To make this representation more useful, we need to extend it to cover polarization. Let $E_x$ and $E_y$ be the two components of the electric field in a quasimonochromatic beam traveling in the $z$-direction. Then, for unpolarized or "natural" light, we can write

$$E_x = a_x(t)\cos\{\omega t - \phi_x(t)\},$$
$$E_y = a_y(t)\cos\{\omega t - \phi_y(t)\}, \tag{11.8}$$

where $a_x$ and $a_y$ are uncorrelated, likewise $\phi_x$ and $\phi_y$.

From equation (11.8) we can describe unpolarized quasimonochromatic light as follows: represent the electric field by a vector in the Argand or phasor diagram; then with time the tip of this vector describes figures ranging from straight lines in different directions through ellipses to circles, and the changes take place at rates comparable to the coherence time of the beam (the coherence length divided by $c$). That is, the unpolarized beam actually changes rapidly and randomly between plane, elliptical, and circular polarization. (The light from a blackbody cavity passed through a suitable narrow-band filter will be quasimonochromatic and completely unpolarized in this sense, but light from most other supposedly unpolarized sources is partially polarized. Thus light emitted from a smooth tungsten filament is partly polarized except for that portion which is emitted precisely normal to the surface. An "unpolarized" HeNe laser usually produces light alternately polarized in two orthogonal directions determined by minute asymmetries in the geometry, the alternation being at a rate dependent on the gain for these two modes.) The above description can clarify the somewhat confusing statements sometimes found in texts to the effect that orthogonally polarized beams do not interfere with one another. Taking "unpolarized" to apply strictly to beams described by equations (11.8), we could first select two orthogonally polarized beams by means of a double-image prism or a polarizing beam-splitter (figure 10.10). From equations (11.8) we know they are uncorrelated and cannot interfere on any observable time scale. Next we could polarize the beam at an azimuth of $45°$ before it entered the beam-splitter by means of a polarizing filter. The two emerging beams would be polarized orthogonally, but they would be completely correlated and would thus interfere. But fringes would be detected not as variations of intensity but as variations of the state of polarization of the combined beams between linear and circular. Finally, if the beam overlap region is observed through another polarizer (analyzer in the usual terminology) oriented parallel to the first, intensity fringes can be seen. Many technical interferometers—in particular, interference microscopes—are based on polarization beam-splitting since the requirements for mechanical stability are much less critical than for interferometers of the Michelson type. A good survey is given by Françon and Mallick (1971).

# 12

# Detectors and Light Sources

## 12.1 Collecting an Optical Image

In trying to pick up an optical image, to measure the intensity of a spectrum line, to detect an interference fringe, to measure the angle of polarization of a light beam, or to obtain any of the other measurements which must be made from the output of an optical system, we are ultimately limited by noise and we have to arrange matters so that the measurement is made with an adequate signal-to-noise ratio. There are two main sources of noise: (a) photon noise, due to the random photon annihilations which produce charge carriers in a photoelectric detector, and (b) detector noise, which arises in a variety of ways in electronic detection systems. To these may be added several other sources, which may or may not be relevant depending on the application: mechanical noise from vibrations in the equipment, noise due to refractive index fluctuations in air paths (of great importance in ground-based astronomy), noise due to thermal drifts in mountings, etc. There is also a sort of noise, not time-dependent, due to optical defects such as bubbles, striae, or stones inside lenses and scratches on the lenses. This sort of noise is more of a nuisance in coherent systems such as interferometers, but a scratch on a surface which is near a real image can contribute to this kind of noise.

It is assumed that in an image-forming system there is adequate resolving power or MTF for the detail which has to be detected behind the noise. (Strictly speaking, if the point spread function is known accurately enough, a deconvolution process can yield a lot of information about detail below the resolution limit if the noise is negligible, but in practice this is very difficult and time-consuming with an image more complicated than just a few isolated bright points.) Then it is found that, just as in communication systems, there is a trade-off, in that the signal-to-noise ratio can be improved both by observing for a longer time and by narrowing the temporal bandwidth used. Of course, while this can sometimes be done in laboratory experiments, it is often not practicable in observational work such as astronomy.

In this chapter we shall be concerned with the operating characteristics such as temporal frequency response, spectral range, and physical size which determine signal-to-noise ratio but we shall not go into the details of the mode of action of detectors such as the structure of photocathodes, the type of doping, and so on. The technology of detectors, particularly of semiconductor devices, is advancing so rapidly that manufacturers' catalog material must be consulted for the latest information.

## 12.2  Classification of Detectors

We may classify detectors in several different ways, and the boundaries will overlap. The main classes refer to spectral range, temporal frequency response, noise limits, size, and whether total flux or an image is picked up. Restricting the discussion to wavelengths from the near-ultraviolet to the thermal infrared, say, 0.2 $\mu$m to 15 $\mu$m, there are four main types of detector:

(a) External photoelectric effect detectors, a classification which includes vacuum photodiodes and photomultipliers.

(b) Internal photoelectric effect detectors, a classification covering a very wide range of semiconductor devices in which photons are absorbed to produce charge carriers.

(c) Thermal detectors, a classification in which it is the direct heating effect of absorbed radiation which is used, e.g., thermocouples, bolometers (resistance thermometers) and the Golay cell, in which the expansion of gas heated by radiation is detected.

(d) Detectors in which a chemical change is initiated by the radiation; the obvious examples are the photographic process and the eye.

### 12.2.1  Detectors Based on the External Photoelectric Effect

The external photoelectric effect can be used from 0.2 $\mu$m to about 1 $\mu$m (in fact almost any material will produce photoelectrons in a vacuum from radiation of wavelengths below 0.2 $\mu$m, but we are restricting our discussion to wavelengths longer than this). The design and manufacture of photocathodes becomes more critical at the longer wavelengths, and different photocathodes are needed for different ranges of wavelengths (also, of course, different windows to the vacuum envelope). The quantum efficiency is the best measure of sensitivity for vacuum photodiodes and photomultipliers. It is the reciprocal of the average number of photon annihilations required to produce one photoelectron, usually expressed as a percentage, and it is given as a function of wavelength in manufacturers' data books. It is between 5% and 30% for most photocathodes at their most efficient wavelengths. Data books also quote sensitivity as amperes of current output per watt of light (or possibly even per lumen, but that is becoming rare except for TV cameras and the like); the conversion to quantum efficiency is trivial.

The response time or rise time of a vacuum photocathode is of order 1 ns, i.e., the frequency response cutoff is beyond 1 GHz. This performance is reduced somewhat with photomultipliers depending on the conditions of use.

Photocathode sizes in photomultipliers vary from about 3 mm to 150 mm. The latter are used for such purposes as scintillation counting where the source is of large size and solid angle, the former for applications where dark current must be minimized.

A vacuum photodiode is in one sense almost an ideal detector: the only internal source of noise is in its dark current, which may be made as low as $10^{-12}$ A by careful design and selection and may be reduced further by cooling. The noise in the signal is ultimately limited to what is usually called shot noise, the random emission of photoelectrons corresponding to random photon annihilation. Shot noise obeys Poisson statistics: if $n_0$ is the average number of photoelectrons per second, then the variance is $n_0$ per second.

Photomultipliers also have shot noise from the electron multiplier stages, but this noise is usually small compared with the effects from the photocathode. Thus the signal strength and the quantum efficiency are the main parameters which determine the signal-to-noise ratio for photomultipliers and photodiodes.

There is a fundamental difference between vacuum photodiodes and photomultipliers in that, if the signal from a photodiode has to be amplified, the amplifier produces further noise (see section 12.3) which may swamp the shot noise from the signal, whereas the electron multiplier stages in a photomultiplier can produce current amplification of order $10^6$ with quite low extra noise. Thus photomultipliers are used for small signals, and the vacuum photodiode finds its main scientific application in measuring very short but intense light pulses, e.g., from pulsed lasers.

### 12.2.2 Semiconductor Detectors

Depending on the material and the doping, semiconductor detectors can cover the spectral range from about 0.5 $\mu$m to 15 $\mu$m (or even further, but this is beyond the range chosen for our discussion). A great variety of semiconductor detectors are available, and as mentioned above it is essential to contact manufacturers for the latest performance figures. Here we shall mention the main types only, giving typical performance under optimum conditions, i.e., cooled when necessary and with appropriate circuits. The quantum efficiency of semiconductor detectors is generally higher than that of external photodevices, but the efficiency is not usually quoted because the noise is not usually dominated by signal noise as it is in photomultipliers. A semiconductor detector is an electrical resistance, and the thermal fluctuation in the current carriers generates a randomly varying voltage which is the main source of noise. Known as Johnson noise or thermal noise, it is present in all resistors as a randomly varying voltage across the terminals. Thus, instead of quoting quantum efficiency for semiconductor detectors, we use the *noise equivalent power* (NEP) or some equivalent. The NEP is the input signal power for which the signal-to-noise ratio is unity, taken per unit frequency bandwidth, and is of order $10^{-10}$ to $10^{-16}$ W, depending on the detector, the wavelength, and the mode of operation. Table 12.1 gives a few of the semiconductor materials together with their properties as detectors. However, this information should be supplemented with manufacturers' technical details, since the performance of the materials varies widely with the type of doping and in some cases with the method of use, i.e., whether as a photovoltaic or a photoconductive device. Also, most detectors used at wavelengths longer than about 2 $\mu$m require cooling, usually to liquid nitrogen temperature, 77 K, to give their lowest NEP. Semiconductor detectors are generally small in area, sometimes less than 1 mm$^2$, since this reduces the Johnson noise.

## TABLE 12.1
### Semiconductor Detectors

| Material | Wavelength Range $\mu$m | Response Time | NEP (watts) |
|----------|------------------------|---------------|-------------|
| Si       | 0.2 – 1.2              | 0.5 ns        | $10^{-16}$  |
| Ge       | 0.5 – 1.8              | 0.3 ns        | $10^{-14}$  |
| InSb     | 1. – 5.5               | 100  ns       | $10^{-17}$  |
| PbS      | 1. – 4 &#124;          | 0.1 ms        | $10^{-12}$  |
| HgCdTe * | 2. – 15 &#124;         | 1  ns         | $10^{-16}$  |

* Different compositions of HgCdTe cover different portions of the wavelength range shown.

Semiconductor detectors for the visible and near-infrared can be obtained in one- or two-dimensional arrays so that extended images can be picked up without mechanical scanning. The elements range from about 10 $\mu$m to 50 $\mu$m in size and up to about 500 by 500 in two-dimensional arrays so that they can be used in TV cameras instead of the older vacuum-tube detectors. These devices are usually operated in a charge-coupled mode (CCD, or charge-coupled device) in which the charges generated by the radiation are read off sequentially at the end of a row of elements.

The devices have the great advantage that the position of each detector element is known and fixed, allowing accurate digitized metrology. The one-dimensional arrays are available with up to 4,096 elements at the time of writing. Also, the individual detector elements can be made with greater extension perpendicular to the scan direction than the detector spacing, an advantage for such applications as spectroscopy. CCDs are usually intended to operate at or near TV rates (exposure time per element of order 0.2 $\mu$s); they can be operated much slower than this, but if they are, problems with charge leakage between adjacent elements can occur.

### 12.2.3 Thermal Detectors

Thermocouples, bolometers, and Golay cells are all very slow in response compared to the semiconductors listed in table 12.1, but the former have advantages for certain purposes. First, they respond like blackbodies to the extent that their radiation-absorbing surfaces can be made black over the required spectral range, i.e., there is no spectral sensitivity calibration. Second, they can be used at room temperature (although in some cases the signal-to-noise ratio is improved by judicious cooling). Thus, although the response time may be as long as 0.02 s in the case of a Golay cell, say, these devices can find their uses in some laboratory experimental work (as opposed to observational work) where it is not necessary to work fast because the phenomena are not changing in time.

### 12.2.4 Photochemical Detection

The photographic emulsion is the prime example of photochemical detection. It is very nonlinear in response, has very low quantum efficiency (although there is no clear agreement about how this should be defined), and does not work in "real time" (although some "instant" photography systems need as little as a few seconds for development). To balance these disadvantages, it has the following great advantages: complete images can be recorded with orders-of-magnitude more pixels (picture elements) than any TV system, and it has the integration property that faint images can be recorded and summed over several hours if necessary, as is routinely done in astronomy. CCD techniques are now used in astronomy for integration over long periods, by storing successive scans, but are still limited to a relatively small number of pixels so that the photographic emulsion still has some uses as a detector for scientific work.

On the other hand the human eye has fewer uses as a prime detector; most *photometry*, radiometry weighted according to the response of the eye, and most color measurements are now done by suitably calibrated photoelectric systems. Nevertheless it is worth noting some of the remarkable properties of the human eye. When adapted to the dark and used suitably, the eye can detect a flash of light containing only a few photons, yet with appropriate light and dark adaptation the eye can be used over perhaps 14 orders of magnitude of intensity. The nominal resolution limit is said to be about 1 arcminute of angle, yet in vernier mode (detecting a transverse break in a line) a few arcseconds can be detected. The available angular field of view in some azimuths is more than $90°$ from the center. The adaptation to different light levels is partly done by the iris, seen from the outside world as the pupil, and this image has considerable extent even when viewed at $90°$ to the axial direction, an effect copied in the design of so-called fish-eye camera lenses.

## 12.3 Noise

We have mentioned shot noise as inherent in the signal and thermal noise generated by random movement of charge carriers in conductors. These are by no means the only sources of noise in detection systems, but they will serve to illustrate the principles behind the reduction of noise effects. A signal is processed electronically with a certain temporal bandwidth which depends on the properties of the detector and the circuits associated with it. Let this bandwidth be $\delta f$ Hz. Then the signal-to-noise ratio if shot noise only were present would be

$$S/N = \{i/(2e\delta f)\}^{1/2}, \tag{12.1}$$

where $i$ is the photoelectron current and $e$ is the electronic charge. This formula illustrates the general principle that with shot noise the signal-to-noise ratio is improved by (a) increasing the signal and (b) decreasing the bandwidth. It is not always possible to increase the signal, but the bandwidth may be decreased by various devices, some of them involving taking a longer time doing the measurement (see section 12.3.1).

The corresponding formula for Johnson noise is

$$S/N = \{R/(4kT\delta f)\}^{1/2}, \tag{12.2}$$

where $R$ is the value of the resistance in which the noise is generated, $k$ is Boltzmann's constant, and $T$ is the absolute temperature. Again, it is desirable to decrease the bandwidth to improve the signal-to-noise ratio, and it can also be seen from this equation why it is necessary to cool most of the detectors listed in table 12.1. It is not merely solid-state detectors which are a source of Johnson noise; it arises in any resistor, but usually only resistors immediately following the detector are important in this connection since their noise contribution is added before the signal is amplified.

### 12.3.1 Improving the Signal-to-Noise Ratio

In this section we mention some special techniques for reducing the effects of noise.

The main source of noise in photomultipliers is shot noise from the signal. The multiplier section constitutes an almost noise-free dc amplifier, but there is a small contribution in the form of shot noise from secondary emission in the multiplier dynodes. In addition, there is shot noise from dark current from the photocathode and there are contributions from traces of natural radioactivity in the glass envelope of the photomultiplier and from cosmic rays. The dark current may be reduced by judicious cooling, since it is partly thermal in origin, and by buying more sophisticated photomultipliers. Then the best way to measure very small signals with a photomultiplier is by the technique of *photon counting*. Thus, if the signal is so weak that there are, say, only 100 photon annihilations per second, each photoelectron produces a pulse of perhaps $10^6$ electrons at the output of the photomultiplier and these pulses can be counted individually instead of trying to measure a very weak current. The average over whatever time is needed to get a small enough variance in the pulse rate is then the required measure of the signal. If the signal is known to be steady, the signal-to-noise ratio can in principle be improved indefinitely by counting for a long enough time.

The technique of *chopping* the signal to impose a carrier frequency on it is useful in many circumstances. The signal is interrupted by, e.g., a sector disk for low frequency or an electro-optic shutter for high frequencies, and the detector circuit has a narrow-band filter tuned to the chopping frequency. The bandwidth $\delta f$ of equations (12.1) and (12.2) is then the bandwidth of this filter, and this is arranged to correspond to whatever bandwidth is expected in the signal. The technique of chopping is used in, e.g., spectrophotometers, where the measurement being made is effectively dc, but it is more convenient to have ac circuits.

A development of ordinary chopping is known by various names, such as *synchronous rectification*, and *homodyne detection*. The chopped signal, (a) in figure 12.1, is amplified and low-pass ac filtered, as at (b), and it is then rectified every half-wave in synchrony with the chopper, as at (c). This signal is then measured with a dc detector with a time constant, say, which is much longer than the chopper period (in effect, the alternating part is filtered out). Then the bandwidth is, apart from a numerical constant close to unity, $\delta f = 1/\tau$, and since $\tau$ can be made very large, the bandwidth can be very small. For example, in measuring small steady infrared signals with a Golay cell and homodyne detection, integration times $\tau$ of several minutes might be used.

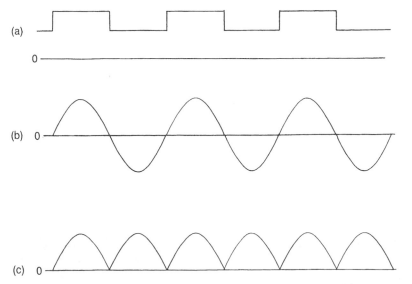

12.1 The principle of synchronous rectification or homodyne detection. (a) A chopped dc or slowly varying signal with a dc component. (b) The signal low-pass amplified with ac coupling. (c) The signal rectified every other half-wave. The magnitude of the original chopped signal is then proportional to the dc level in (c) but with reduced bandwidth as explained in the text.

## 12.4 Radiation Sources

### 12.4.1 Visible Light Sources

Visible light has its own system of radiometry, called photometry, and the radiometric quantities described in section 8.1 have their equivalents in photometry. The basic unit is again that of flux, the *lumen*. Elaborate experimental techniques have been devised to determine the equivalent in lumens of a watt of light power at different wavelengths, and the results are expressed by the *relative visibility curve* of figure 12.2. To get an absolute scaling, the value at the wavelength of peak visibility, 555 nm, is 683 lumens per watt. (The lumen was originally defined in terms of the light output of a "standard candle," but nowadays the lumen is actually *defined* by the ratio 683 lumens per watt at 555 nm together with the data in figure 12.2.) The other quantities corresponding to those in section 8.1 are illuminance, which is lumens per unit area falling on a defined surface, and luminance, which is lumens per unit solid angle per unit projected area from a light source.

Filament lamps are the most common form of thermal or incoherent

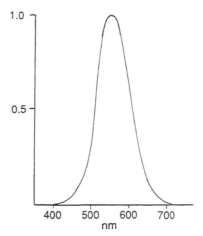

12.2 The luminous efficiency of the human eye (photopic or bright conditions). The ordinate represents the relative impression of brightness of radiation of constant intensity per unit wavelength interval.

source for visible light, and a tungsten filament has to a good approximation the radiation emission characteristics of a *gray body*; i.e., the spectrum is like that of a blackbody radiator but with all intensities scaled down by the same factor $\epsilon$, which for tungsten is of order 0.5. This factor is, of course, the emissivity.

Quartz-halogen filament lamps can be run at color temperatures up to 3,300 K. Under these conditions the luminous efficiency is about 25 lumens per watt of electrical power used in the lamp. (This is much less than the figure 683 lumens per watt mentioned above, partly because not all the electrical power goes into generating radiation and partly because, from figure 12.2, all visible wavelengths other than 555 nm are less visually efficient than 683 lumens per watt. The "quartz" in quartz-halogen is actually fused silica, not crystal quartz.)

The luminance of a tungsten filament at 3,300 K is about $3.10^7$ lm sr$^{-1}$ m$^{-2}$. Sources with greater luminance are, for example, high-pressure mercury arcs, $4 \times 10^8$; xenon arcs, $10^9$; and metal-halide arcs, $10^9$ lm sr$^{-1}$ m$^{-2}$. These sources have color temperatures up to about 6,000 K (but "color temperature" is approximate in this context since the spectral distributions do not precisely match those of blackbodies at any temperature).

The concepts of luminance and radiance as discussed here belong to geometrical optics and extended incoherent sources. Thus they cannot be directly applied to a single transverse-mode laser such as the ordinary HeNe laser at 632.8 nm, since this is a coherent source and in the geometrical optics model a Gaussian laser beam as described in chapter 9 comes from a single point source of zero area. (At any given point $P$ along the Gaussian beam the phasefront has a certain radius of curvature, $R$, in the notation of chap. 9, and the point source for that region of the beam would be the corresponding center of curvature at a distance $R$ from $P$. This rather unrealistic statement is made to emphasize the fact that we need to exercise care when talking about lumens or watts per unit area per unit solid angle from a single-mode laser beam.)

The radiance or luminance for single-mode lasers can be introduced in a consistent way by assigning to the expression given in equation (8.2) the value $B\lambda^2$. That this is a reasonable assignment follows from taking $\pi w_0^2$ for the area of the Gaussian beam waist, and equation (9.5) for the divergence angle $\alpha \approx \theta$. Therefore, we may ascribe to a single-mode laser the radiance

$$B = \text{power}/\lambda^2. \tag{12.3}$$

As an example consider a 1 mW single-mode HeNe laser operating at a wavelength $\lambda = 632$ nm. According to our formula, the radiance is 2,500 w/mm$^2$. This example serves to illustrate that even a low-powered laser is a very bright source indeed.

A comparison of lasers and thermal sources is interesting in other ways. Using very approximate figures, a quartz-halogen lamp emits perhaps 10% of its input power as visible light, spread over the visible spectrum and over solid angle $4\pi$, so we get a total of, say, 10 watts of light from a 100 W lamp compared to 1 mW from the above-mentioned laser. However, in a bandwidth of 0.1 nm (still much larger than that of the laser) and over $10^{-6}$ steradians (roughly the divergence angle of a 1 mm diameter laser beam), the filament radiates only about $10^{-12}$ W mm$^{-2}$. Thus it is the directionality of the laser beam rather than total power which makes it appear much brighter.

## 12.4.2 Infrared Sources

Apart from lasers, most laboratory sources of the infrared are blackbodies of various forms. The blackbody cavity is used for precision laboratory and standards work since its spectrum follows the Planck law and if properly designed it is a Lambertian radiator: the radiance depends only on one parameter, the temperature. More convenient sources are variations on and developments of the Nernst glower: a rod formed from a mixture of rare-earth oxides and heated electrically to run at between, say, 1,500 K and 2,000 K, depending on the spectral region of interest. A Globar is a similar structure but made of silicon carbide. A tungsten filament is also a source of infrared, but it is useless at wavelengths longer than about 2.5 $\mu$m on account of the absorption of the glass or fused silica envelope. Globars and similar sources are less susceptible to vibrations than tungsten filaments, as

the former are usually in the form of a tube or rod a few millimeters thick;
they appear to behave as Lambertian sources to a good approximation.

# 13

# Image Scanning and Beam Deflection

## 13.1 Introduction

Direct image formation by lenses and mirrors has reached a highly developed stage, both in the design of optical systems and in detection systems, but there are many purposes for which alternative methods of image capture are preferable. Consider, for example, the automated imaging of large printed circuit boards, such as are used in desktop computers, for assembly of components on the boards or for checking the correct positioning of components on them. To get enough detail in the image, a view of perhaps 5,000 or 10,000 pixels across the width of the board may be needed, but this is beyond the range of any TV tube or two-dimensional CCD array. In a *scanning* system a single spot of light is focused in the region of the board or other object and is made to scan, usually in a rectangular raster, by mechanical or other scanning means. Light scattered from the spot is picked up by a photodetector, and the image can be reconstructed or otherwise processed as desired.

Let an untruncated TEM$_{00}$ laser beam be focused with its waist at the level of the board or other object. Then from the equations of section 9.1 we can obtain an expression for the depth of focus of such a scanning system, for, defining the extremes of the range of focus as those points at which the intensity in the center of the spot drops to 80% of its value at the waist, the focal range is

$$\Delta z = \pm 0.45\lambda/(\pi\theta^2) = \pm 0.45\pi\omega_0^2/\lambda, \qquad (13.1)$$

where $\theta$ is the asymptotic semiangle of the beam (eqn. 9.6). A typical value for $\omega_0$ would be 25 $\mu$m in such an application, and with $\lambda$ in the visible region this leads to $\Delta z$ of order $\pm 1.5$ mm.

## 13.2 Mechanical Scanning

Most mechanical scanning systems use one of three methods of scanning the laser beam: (a) oscillating mirrors ("galvanometer mirror"), (b) rotating mirrors, usually in the form of a polygonal block with mirror faces on each face, and (c) rotating holograms. A good survey with much engineering detail is given by Marshall (1985). In this section we discuss some of the optical design points of mirror and polygon scanners. We defer holographic scanners to Chapter 15.

A one-dimensional or line scan may be obtained with the lens that produces the beam waist either before or after the scanning mirror, as in figure 13.1 (a) and (b), respectively. With postlens scanning, as in (a), the lens operates only on its axis and can therefore be of very simple design. But it can be seen that the beam waist travels in a circular arc, and thus there is a depth-of-focus problem for a wide scan. This has been overcome in different ways, including dynamic focusing of the lens and auxiliary field-flattening elements after the scanning mirror. With prelens scanning, as at (b), the lens must cover a wide angular field of view and, what is in practice more of a nuisance in design, its entrance pupil should be outside the lens components, at the scanning mirror. Special lens designs are now available for this kind of scanning geometry. They are sometimes designed with a distortion characteristic that makes the image distance proportional to the field angle so that the scan is linear with the angular displacement of the scan mirror (normal distortion correction requires that the image distance is proportional to the tangent of the field angle). Such lenses are called $f\theta$ lenses since for an object at infinity subtending the angle $\theta$ the image size is $f\theta$. Prelens scanning becomes more difficult with a two-dimensional scan, achieved by two scan mirrors in tandem, since the aperture of the lens gets rapidly larger.

Similar remarks apply to rotating polygon scanners, but the details of the geometry are different. There are intricate trade-offs to be worked out with the duty cycle of each polygon facet versus vignetting of the pupil. Thus it can be seen from figure 13.2 that, if the width of the laser beam corresponds exactly to that of a facet, the next scan will start just as the last one finished (assuming the lens aperture is large enough), so that there is a very good duty cycle, but the full width of the beam is used only at the middle of each scan; sufficiently wider facets eliminate the vignetting but give a worse duty cycle. Marshall (1985) deals in detail with dynamical considerations in running both polygons and mirrors at high speeds. Ready-built units of both types are now commercially available.

Mirror scanners can be used in a random address or vector fashion (provided they are not simply driven in resonant mode), and this can be useful for certain inspection and recognition purposes (random addressing is, of course, not possible with polygon scanners). Polygon scanners are used, for example, in infrared TV systems for the thermal band (8 to 14 $\mu$m) for viewing ambient temperature scenes. For this purpose no two-dimensional detector arrays with enough sensitivity are available and one-dimensional

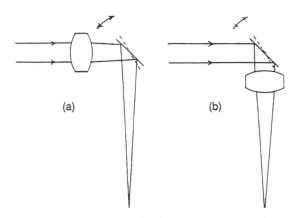

(a)  (b)

13.1 Scanning with an oscillating mirror (galvanometer scanner). (a) Postoptics scan. (b) Preoptics scan.

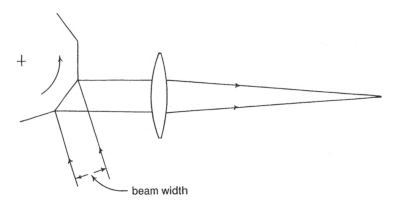

beam width

13.2 A polygon scanner in preoptics scan mode.

arrays have only a small number of elements, so that a polygon scanning system is the natural choice.

## 13.3 Electro-optical Scanning

All dielectrics change in refractive index under an applied electric field so that the deviation produced by a prism can be changed in this way.

For a detailed description with applications see Gottlieb, Ireland, and Ley (1983). The materials used are typically crystals of low symmetry such as $NH_4H_2PO_4$, the field strengths are limited to $10^4$ V mm$^{-1}$, and the maximum index change achievable is of order 0.001. Thus with a single prism the maximum deflection is of order 1 mrad. To get larger deflections, systems with many prisms in tandem have been used.

The performance of such deflectors may be categorized by a parameter $N_r$ which signifies the number of resolvable spots at different angles. This number is obtained by dividing the maximum deflection angle by the diffraction angular width of the light beam, which is equal to the wavelength divided by the width of the beam. Values of $N_r$ for multiple-prism systems in the order of 1,000 are quoted by Gottlieb et al.

Obviously electro-optic deflectors cannot match polygons or mirrors in angular range. Their particular advantages for suitable applications are (a) that there are no mechanical moving parts and (b) that the response time can be short, of order 1 $\mu$s. However, in practical cases this can be limited by the performance of the crystal as a circuit element, i.e., its capacitance and resistance: the former leads to longer time constants and the latter to ohmic heating.

## 13.4 Acousto-optic Deflectors

Mechanical stress—compressive, tensile, or shear—changes the refractive index of dielectrics; this phenomenon is called the photoelastic effect, and it leads to the formation of, in effect, a diffraction grating in a medium through which a sound wave is traveling (Gottlieb et al. 1983). Figure 13.3 depicts a sound wave driven through a crystal by a transducer attached to one end. The grating is formed as a distribution of high and low refractive index of sinusoidal form, or a *phase grating*, and it travels at the speed of sound in the medium. The fact that it is a moving grating has no effect on its action as a transmission diffraction grating except to change slightly the frequency of the diffracted light by the Doppler effect.

Diffraction can take place in two different modes, depending on the index modulation and the optical thickness in the direction of travel of the light. For small modulation and small optical thickness many diffracted orders are produced, as in, say, a conventional diffraction grating consisting of equal opaque and transparent portions; this is the Raman-Nath mode. For use as a deflector the index modulation and optical thickness are greater and then the device operates in the Bragg mode; that is, only one diffracted beam is produced, the diffraction efficiency is high, and the equivalent of the Bragg condition for X-ray diffraction in crystals must be satisfied:

$$2\Lambda \cos \theta = \lambda, \tag{13.2}$$

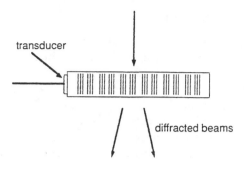

13.3 Diffraction by an optoacoustic modulator.

where $\Lambda$ is the acoustic wavelength and $\theta$ is the internal angle of incidence, as in figure 13.4. (Eqn. 13.2 applies to isotropic media; the effect is different in anisotropic media.)

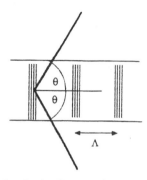

13.4 Diffraction in the Bragg regime by an acoustic wave.

Sound speeds in dielectrics are of order two to three km $s^{-1}$ and the transducers are driven at frequencies approaching the GHz range, thus giving acoustic wavelengths in a suitable range for diffracting visible light. The number of different resolvable diffracted directions depends on the size of the device and on the range of acoustic frequencies, but it is also limited by the need to stay near the Bragg configuration in order to preserve diffraction efficiency. Acousto-optic deflectors are therefore small-angle devices, like electro-optic deflectors, and they have their advantages in special niches, some of which are discussed by Gottlieb et al. (1983).

## 13.5 Scanning Microscopes

So far in this chapter we have discussed modes of scanning at relatively low numerical aperture and over large distances. In a very different type of scan technology small distances and large numerical apertures are involved. Figure 13.5 shows the elements of a *scanning microscope*; the principle is that the object is illuminated by a single spot of light small enough to be the point spread function of the objective and the image is built up from a raster scan. The scan can be formed by actually moving the object stage in two dimensions. Doing this requires very good mechanical work but has the advantage that the optical system is always used on axis. Thus a flat-field optical system of field coverage limited only by the stage design is obtained.

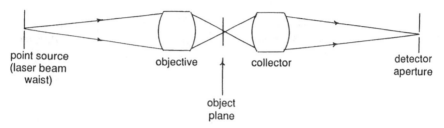

point source (laser beam waist)      objective      collector      detector aperture

object plane

13.5 Optical scanning microscopy. The object is moved in a raster pattern to build up an image.

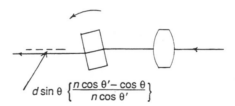

$$d \sin \theta \left\{ \frac{n \cos \theta' - \cos \theta}{n \cos \theta'} \right\}$$

13.6 Displacing a beam laterally by tilting a plane-parallel block of glass. In the formula for the displacement, $\theta$ is the angle of tilt from the normal, $d$ is the thickness of the block, $n$ is its refractive index, and $\theta'$ is given by $n \sin \theta' = \sin \theta$. For most purposes the cosines in the formula can be made equal to unity.

Alternatively, the source could be scanned as in figure 13.6 by a tilting plate mechanism or by galvanometer mirrors as in section 13.2. A third method is to use a CRT raster as the source; here one is limited both in signal strength and in number of pixels, but there are the advantages of high scan speed and no moving parts. If the detector has the same dimensions as the source (allowing for any difference in magnification), then both are in effect point spread functions and the performance is diffraction-limited. This arrangement of source and detector as diffraction-limited points has important advantages in resolution and focal depth, particularly in the microscopy of opaque objects (Sheppard 1984).

## 14

# Diffraction Gratings

## 14.1 Introduction

Diffraction gratings have their principal use as dispersing devices for spectroscopy, but from the point of view of this book they are also useful as a preview to the subject of holography. In this chapter we review without proofs the main properties of gratings and we note some applications.

## 14.2 Properties of Plane Diffraction Gratings

Figure 14.1 shows schematically a plane diffraction grating with ruling spacing (*grating constant*) $\sigma$. A plane wave is incident at angle $\alpha$ and it is diffracted at an angle $\alpha'$, both positive as shown. The *grating equation* is then

$$\sin \alpha + \sin \alpha' = m\lambda/\sigma, \qquad (14.1)$$

where $m$ is the order of diffraction. In this equation it is tacitly assumed that the rulings are perpendicular to the plane of incidence and that we are dealing with a reflection grating; very few spectroscopic gratings are used in transmission. Equation (14.1) gives the direction of the diffracted beams of different orders but does not indicate their relative intensities. The relative intensity distribution depends on the profile of the rulings as well as their spatial frequency (and also on the material, although most gratings are made of aluminum), and when, as is usually the case, $\sigma$ is similar in magnitude to $\lambda$, it is difficult to calculate the relative intensity distribution with any reasonable accuracy. For many purposes it suffices to *blaze* a grating, i.e., to concentrate most of the diffracted light into one order by profiling the rulings as in figure 14.2: the grating profile is approximately of sawtooth shape with the normal to the active face at an angle $\theta$ to the grating normal; $\theta$ is called the blaze angle. Then, if the diffracted spectrum is centered around an angle $\alpha'$ such that

$$\alpha' + \alpha = 2\theta, \qquad (14.2)$$

most of the light will be in this spectrum. The reasoning here is that the direction $\alpha'$ corresponds to the maximum of diffraction by a single slit of width $\sigma$ and this single-slit diffraction pattern is the diffraction envelope governing the distribution among different orders. In practice there is usually less light in this order than would be expected on this simple reasoning, partly because the ruling profile cannot be made precisely enough and partly because the simple scalar diffraction theory on which this reasoning is based is not adequate.

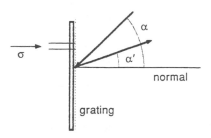

14.1 A plane diffraction grating used in reflection mode.

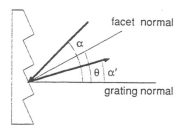

14.2 Blaze angle for a reflection grating.

Corresponding to the point spread function for image-forming systems (section 7.1), a spectroscopic instrument has a *line spread function* which, if there are no aberrations in the system, takes the form of a sinc$^2$ function, i.e., suitably scaled $\{(\sin z)/z\}^2$. The spectroscopic resolving power in a region of the spectrum of wavelength $\lambda$ is then conventionally defined as $\lambda/\delta\lambda$, where $\delta\lambda$ is the separation between two adjacent wavelengths such that the maximum of the sinc$^2$ of one falls on the first minimum of the other. Then for a grating containing $N$ rulings the resolving power is

$$\lambda/\delta\lambda = mN, \qquad (14.3)$$

where, as in equation (14.1), $m$ is the order of diffraction.

A basic equation for plane gratings refers to overlapping orders. One important difference between prisms and gratings as spectroscopic devices is that prisms produce only one spectrum whereas, as can be seen from equation 14.1, gratings produce spectra of different orders. Explicitly, for a given angle of incidence, wavelength $\lambda_2$ in order $m + 1$ will appear at the same position as $\lambda_1$ in order $m$ if

$$(m + 1)\lambda_2 = m\lambda_1. \tag{14.4}$$

From this it follows that in the region of wavelength $\lambda$ and in order $m$ the *free spectral range* is

$$\Delta\lambda = \lambda/m. \tag{14.5}$$

As with prisms, a spectrum line formed by a plane grating is curved (recall eqns. 4.5, 4.6) because of the skew incidence of rays not from the center of the entrance slit of the spectroscope. The expression for the curvature is

$$m\lambda/\{f\sigma \cos \alpha'\}, \tag{14.6}$$

where $f$ is the focal length of the objective which brings the spectrum to a focus and the other symbols have the same meanings as in the other equations in this section. The curve is concave toward the longer wavelengths, the opposite sense from the curvature due to prisms.

Equation (14.6) is the last of the basic formulas for plane gratings. Spectroscopic systems involve collimators and objectives for focusing a spectrum, and these are designed in detail by tracing rays through the system. Perhaps paradoxically, rays *are* traced "through" diffraction gratings (although they are usually used in reflection as in our figures) in spite of the fact that they operate by diffraction, so that spot diagrams and other aids to image assessment can be obtained. Routines for raytracing through gratings form part of most of the software packages mentioned in section 6.7.

Another kind of application of plane gratings is for wavelength selection in, e.g., gas ion lasers, which can give several different wavelengths. The grating is used instead of one of the mirrors in the resonator and is set so that the desired wavelength has equal angles of incidence and diffraction (the so-called Littrow geometry as in figure 14.3). Alternatively, the grating may be used outside the laser cavity, but then it is necessary to ensure that the beam is wide enough across the grating to give enough resolving power to separate the desired wavelength from the next nearest of those produced by the laser.

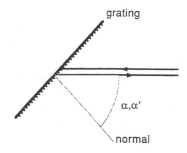

14.3 A plane grating in Littrow geometry.

## 14.3 Concave Gratings

The concave diffraction grating is a hybrid device which both disperses and focuses, and as might be expected the result is a compromise in which neither function is performed as well as it would be by a plane grating with good focusing optics. Nevertheless there are at least two special niches where the concave grating comes into its own, so we give its main properties here. The classical concave grating has rulings following the intersections of equispaced parallel planes with the surface of a concave spherical mirror (holographic concave gratings will be dealt with in chap. 15).

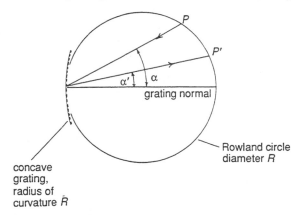

14.4 A concave diffraction grating and its Rowland circle. A source at any point $P$ on the Rowland circle is imaged by diffracted rays in the plane of the diagram at another point $P'$ on the circle. The respective directions of the rays are linked by the grating equation, eqn. (14.1). There is heavy astigmatism; i.e., rays not in the plane of the diagram do not form a sharp focus at $P'$.

Figure 14.4 shows a concave grating with radius of curvature $R$ and spacing $\sigma$. A ray with angle of incidence $\alpha$ is diffracted at the angle $\alpha'$ according to the grating equation (eqn. 14.1), but we have to ask what happens to a pencil of rays from, say, the point $P$ on the incident ray. We expect these rays to form a focus at some point $P'$ on the diffracted ray, and while this is true in principle, in general the focus is very badly aberrated and the spectrum is not well resolved. It was shown by H. A. Rowland that, if $P$ lies on the circle shown of diameter $R$, then $P'$ also lies on this circle for rays in the plane of the diagram. There is still some aberration, but it is mainly a very large astigmatism, and this is sometimes not important in spectroscopy when an image of a slit is being formed. Thus the Rowland circle is a good locus on which to base mountings for concave gratings. One of the main applications nowadays is in vacuum ultraviolet spectroscopy, where all metals have low reflectivity and the gain over plane grating systems is that only one reflecting surface is needed (the grating itself) instead of three. Then it often happens that $\sigma$ is much larger than $\lambda$, and this leads to the grazing incidence type of mounting (figure 14.5), where the large angles of incidence and diffraction help to keep up the reflectivity of the grating surface.

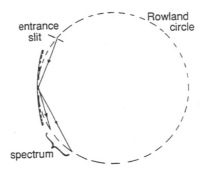

14.5 Grazing incidence geometry on the Rowland circle.

The other notable application of the concave grating does not use the Rowland circle, or, at any rate, it starts at the Rowland circle and moves away from it. There is a need, again often in vacuum uv work, for a compact monochromator, a system in which wavelengths are selected simply by rotating the grating and the different wavelengths emerge in a fixed direction from a fixed exit slit. It was found by M. Seya that, if a concave grating is used as in figure 14.6, with $\alpha = -\alpha' \approx 35°$, then, as the grating is rotated as indicated, the increase of aberrations on departing from the Rowland circle is minimal. Of course, at the position shown in the figure it is the zero order (undispersed light) which emerges from the exit slit, so some nice optimizing of the chosen angles at zero order is needed, depending on the angular aperture of the grating used and on the spectral range to be covered. The Seya mounting is now used widely for compact monochromators of moderate resolution when the materials problems mentioned above arise.

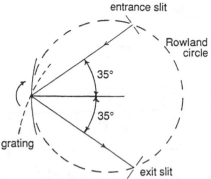

14.6 The Seya-Namioka mounting for a monochromator with a concave grating. The grating is rotated as indicated to scan through the spectrum, and the geometry gradually moves away from the Rowland circle.

A survey of these and other classical grating mountings with details of aberrations is given by Welford (1964). A recent review of exact diffraction theory for gratings was given by Petit (1980).

# 15

# Some Applications of Holography

## 15.1 The Principle of Holography

Let two coherent collimated beams fall on a screen at angles of incidence $\alpha$ and $\alpha'$ as in figure 15.1. A set of straight interference fringes is formed on the screen, and it is easily shown that the spatial period of the fringe system is

$$\sigma = \lambda\{\sin\alpha' - \sin\alpha\}^{-1}. \tag{15.1}$$

If the screen is a photosensitive surface such as a photographic emulsion, the fringe system can be recorded and will appear as a diffraction grating of grating constant $\sigma$ and with an approximately sinusoidal profile of transmission. If this grating is now illuminated by one of the collimated beams, say, that at angle $\alpha$, it will, by equation (14.1), produce a diffracted phasefront in the direction $\alpha'$ as well as other diffracted orders. Ignoring for the moment the other orders, in the language of holography we have recorded a hologram of the $\alpha'$ beam using the $\alpha$ beam as a *reference beam*, and we have reconstructed the $\alpha'$ phasefront (the *object beam*) using the $\alpha$ beam as a *playback* or *reconstruction* beam.

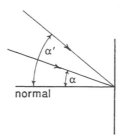

15.1 Two interfering beams forming straight fringes on a photographic plate or other detecting device.

It is then plausible to suggest that, if the $\alpha'$ beam is not collimated (if it comes from a source point at a finite distance), it will still be reconstructed and also that the process will still work if the reference beam is noncollimated. This is all confirmed by more detailed analysis and by experiment. Furthermore, the object beam can comprise any number of individual point sources or a continuum of them, provided object and reference beams are coherent, and a reconstruction is obtained, albeit possibly confused by speckle and multiple order reconstructions of the original.

## 15.2 Holographic Recording Materials

Photographic emulsion was the first material to be used for recording holograms, but it has the disadvantages that the developed emulsion grains are of finite size and that the hologram appears as variations of absorption; this makes the process very lossy: the theoretical maximum efficiency from reconstruction beam to the image is only 6.25%, and in practice it is much less than this. The grain size produces speckle, which obscures the detail in the reconstructed image. The absorption loss is overcome by bleaching the hologram to leave clear regions of varying refractive index so that the developed hologram is a phase structure like the ultrasonic waves of section 13.4 but, of course, not moving. This leads to the distinction between *thick* and *thin* holograms: a thin hologram can produce reconstructed images in several orders, rather as in the Raman-Nath mode of ultrasound diffraction, whereas a thick hologram, which has to be a phase structure, operates in Bragg mode and can have an efficiency approaching 100%.

Many other media have been used for holographic recording, but we mention only those appropriate to the applications to be discussed in section 15.3. *Photoresists* can be deposited as thin layers on glass or other substrates. On exposure to light a latent image is formed, and in the development process the resist is removed in thickness proportional to the exposure (this would be a positive resist; in negative resists the unexposed regions are removed on development). Thus a photoresist can form a phase hologram, but the phase structure appears as a variation of geometrical thickness.

*Dichromated gelatin* (DCG) is just what its name says, i.e., an aqueous solution of natural gelatin treated with potassium or ammonium dichromate and spread as a thin film. When it is exposed and developed, refractive index changes occur which follow the exposure pattern, and so again a phase hologram is obtained. DCG, like photoresists, is in effect grainless, but unlike photoresists it can be coated thickly enough to form thick holograms with very high efficiency.

For other holographic materials and for practical details of techniques in holography see Hariharan (1984) and Smith (1977).

## 15.3  Applications of Holography

Leaving aside displays (such as holographic "art") and security (credit card holograms and the like), there are some technological applications of holography and we shall describe four of them: (a) diffraction gratings, (b) testing non-null phasefronts, (c) holographic optical elements, and (d) holographic interferometry for nondestructive testing.

### 15.3.1  Holographic Diffraction Gratings

The principle of the formation of a diffraction grating was discussed in section 15.1.  To get a usable plane grating such as was discussed in chapter 14, it is necessary to get the profile right for blaze and much of the manufacturers' effort is no doubt devoted to this.  From section 15.2 it is clear that the appropriate material for making a grating is positive photoresist coated afterward with aluminum; many copies of a good master can then be made by replication in plastics.  The elaborate techniques developed for classic grating ruling engines to keep the rulings uniform are all needless because the spacing is automatically held correct in terms of the wavelength of the light and the geometry of the interference system according to equation (15.1).

The holographic method can also be applied to making concave gratings.  Figure 15.2 shows how the two interfering beams might be arranged to produce a Rowland circle design such that, in the exposure wavelength, light from $P_1$ is diffracted in first order to $P_2$.  The rulings will not, of course, be the intersections of parallel equidistant planes with the grating surface as described in section 14.3; in fact their shapes and spacing will automatically be such that the image formation from $P_1$ to $P_2$, will be aberration-free, whereas for the classic concave grating there would be heavy astigmatism.  The property that the diffracted image is aberration-free is very attractive, but it is unfortunate that it does not hold for other wavelengths diffracted from $P_1$ or for diffraction of the original wavelength from any other point than $P_1$.  However, it is possible to optimize the geometry for a given wavelength range and resolution to get considerably better performance than with a classically ruled concave grating.  The most up-to-date information is best obtained from manufacturers.

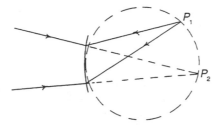

15.2 Forming a Rowland circle concave grating holographically.

## 15.3.2 Testing Non-null Phasefronts

A null test is one in which the component or system under test should show zero aberration. A lens component on its own may not give an aberration-free image and may not be intended to do so, yet for adequate reasons one may be required to test the component on its own. While it is relatively easy to judge whether the performance of a system is aberration-free to within some tolerance (null test), it may be difficult to tell whether the desired large aberration is present in such a component, since it cannot be null-tested on its own; an extreme example would be the corrector plate in the Schmidt camera (figure 6.24c). For workshop testing purposes one may attempt to design a simple optical system consisting of one or two lenses with spherical surfaces which can be put together without risk of error to produce a compensating aberration so that together with the component under test a null test is possible. This is not always practicable, but holography provides a solution, as in figure 15.3. The hologram has the property that when played back by the incoming collimated beam the reconstruction is the required aberrated beam, as indicated. This would seem to be no solution, since a suitably aberrated beam is needed to form the hologram, but the hologram is not formed optically at all; it is computer-generated. Knowing the required aberration allows one to compute the shape of the fringes needed, and these are then generated onto a suitable medium. For more details see Birch and Green (1972) and Malacara (1978).

## 15.3.3 Holographic Optical Elements.

We saw in section 15.3.2 how phasefronts of prescribed shape can be generated holographically. It is a small step from this to the idea of holograms as lenses or mirrors (holographic optical elements, or HOEs). Figure 15.4a shows the geometry for forming a hologram of a convergent beam with a collimated reference beam. On playing back with the original collimated beam as in figure 15.4b the convergent beam is obtained, or in effect the collimated beam has been brought to a focus $F$. The hologram has concen-

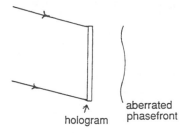

aberrated
phasefront

hologram

15.3 Generating an aberrated phasefront holographically. The hologram would usually be computer generated.

tric circular fringes with radius proportional to the square root of the fringe order, in paraxial approximation; in other words it is like a zone plate.

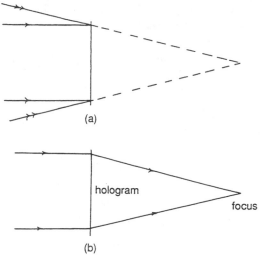

(a)

hologram

focus

(b)

15.4 A holographic optical element ("convex lens"). (a) Forming the hologram. (b) The lens producing a focus.

The HOE reconstructs the phasefront exactly as it was; i.e., there is no aberration for that particular phasefront and in the same wavelength as was used for forming the hologram. However, there is the equivalent of very large chromatic aberration, in the sense that the focal length varies inversely as the wavelength of the radiation, and also at any other wavelength than that used for forming there will be spherical aberration. Furthermore, the HOE as in figure 15.4 has strong off-axis aberrations, i.e., coma and astigmatism. Finally, there is the problem of subsidiary foci from other orders of diffraction in the hologram, unless the diffraction efficiency can be made almost 100%, but with all these difficulties there are still special applications for HOEs. The best-known example is the use of HOEs as focusing reflectors over a narrow wavelength band for head-up displays in aircraft.

### 15.3.4 Holographic Interferometry

Conventional interferometry as described in chapter 11 provides a means of detecting and measuring displacements with a precision of the order of the wavelength of the radiation used, but it is necessary to have an optically smooth surface to measure from and the displacements measured are only the components normal to this surface. Holographic interferometry provides a means of measuring displacements of optically rough surfaces and is not restricted to components of the displacement normal to the surface.

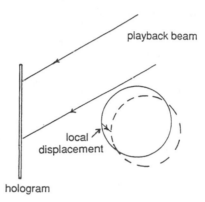

15.5 The principle of holographic interferometry. The fringes map the local displacement between corresponding points in the two positions of the object before and after movement or distortion.

Figure 15.5 depicts the process of making a hologram of an object with a rough surface. The recording medium, say, a photographic plate, is exposed simultaneously to the reference beam and to coherent light scattered off the object. If the object is removed and the hologram is developed and played back, it reconstructs exactly the complex amplitude distribution which had been scattered from the object, which is, of course, why an image of the object is seen when one looks into the hologram. If instead of removing the object it is slightly moved from its original position, as suggested by the broken lines, it will scatter a complex amplitude distribution corresponding to its new position and this will interfere with the reconstructed distribution from the original position. Then on looking into the hologram the object will be seen crossed by interference fringes which map the displacement. The relation between the displacement and the fringes seen is more complicated than in classical interferometry: figure 15.6 shows a point $P$ in its original position and the same point after displacement to $P'$. If the vector displacement from $P$ to $P'$ is $d$ and if $s$ and $s'$ are unit vectors along the illumination and viewing directions, then the path difference which is mapped by the fringes is

$$d \cdot (s' - s), \tag{15.2}$$

from which it can be seen that by this technique even displacements in the plane of the surface may be detected with appropriate geometry of illumination and viewing.

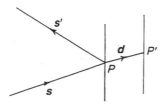

15.6 The fringe function in holographic interferometry. $P$ and $P'$ are corresponding points.

Holographic interferometry is widely used in nondestructive testing, and there are many variations and developments. Up-to-date descriptions with good practical detail are given by, for example, Jones and Wykes (1983) and Gåsvik (1987).

# 16

# Assembling an Experimental Optical

# System

## 16.1 Specifications

The aim of this chapter is to indicate how to go about assembling an optical system for a given experimental purpose. It is desirable to be specific about that purpose because knowing the purpose will indicate which of the matters discussed in the preceding sections should be taken into account. Knowing the purpose may also lead to an operational specification involving, for example, wavelength range, spectral resolution, spatial resolution, time resolution (or speed of response), and contingent things like ruggedness, adverse environmental conditions such as an extreme temperature range, compactness, cost limitation, availability within a certain time, and portability. While it is, of course, impossible to give detailed procedures for even a restricted number of cases, it is useful to consider the kind of questions that can arise in a few instances, and this we shall proceed to do.

## 16.2 An Example of an Imaging System

Forming an image of a transparent object with magnification and resolution and with artificial illumination is a task which occurs in several different forms. The classical problem of microscope design exemplifies this task. Figure 16.1 is a block diagram of the elements of a conventional microscope arranged for *transillumination*, as it is called.

A way to initiate the design of this system is to start with the object of which an image is to be formed and ask what are the expected characteristics. The scale of detail determines the performance of the objective, that is, its numerical aperture. A phase object is one in which the detail is in the form of variations of optical thickness, due possibly to refractive index variations and/or variations in metrical thickness. Pure phase objects do not show up very well in ordinary illumination, and it may be necessary to use interference contrast (coherent illumination with some arrangement of

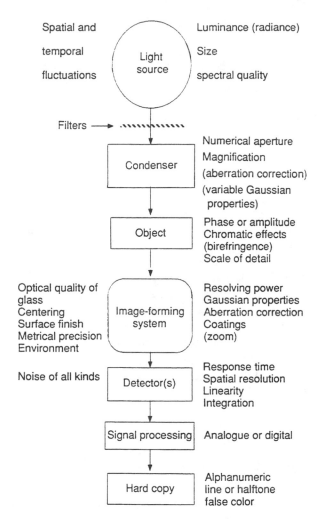

16.1 Schematic of a microscope optical system.

a reference beam unperturbed by the structure in the object and allowed to interfere with the object beam). Since most objects have both phase and amplitude structure, ordinary transillumination may be adequate. If fine shades of color are important, as in stained biological objects, an illumination system well corrected for chromatic aberration is indicated. Thus we are already concerned with details of the illumination system and by implication the light source. If the object is expected to be birefringent, as with petrographic samples, auxiliary polarizing equipment is needed and the objective must be specified to have low birefringence ("strain-free").

We have already discussed condenser systems in some detail in chapter 8, and for flexibility some kind of Köhler illumination system will be needed. The light source must be large enough to allow adequate illumination NA (numerical aperture) and object field size, and it must have enough luminance (radiance) for whatever detection system and bandwidth are planned. The left-hand column in figure 16.1 refers to regrettable defects rather than the intrinsic properties listed on the right; thus fluctuations in the light source must be small enough in magnitude and bandwidth not to interfere with the detection process.

Apart from the requirement of adequate resolution already mentioned, the image-forming system must have a magnification matched to the detector. However, we can stress that the magnification and the resolving power are dependent on quite different attributes of the design: the resolving power depends essentially on the NA of the imaging system, i.e., the size of the cone of rays collected from the object (this ignores effects depending on the coherence of the illumination and the range of wavelengths of radiation), whereas once an image has been collected with adequate NA it is a relatively minor problem to magnify it to whatever extent is needed to match the detector. It is here that the Gaussian properties of the image-forming system as a whole come in. For example, in a microscope used visually the eyepiece acts as a further magnifying stage after the objective to ensure that the detail in the final virtual image seen in the eyepiece is resolvable by the eye (the angular resolution limit of the human eye used under optimal conditions is usually taken to be about one arcminute). But if a CCD array with 13 $\mu$m pixels is used as detector, it could well be possible to dispense with the extra eyepiece magnification, since an objective magnification of, say, $\times 40$ would make the CCD pixels correspond to the resolution limit of an objective of high NA.

The manufacturing requirements of systems like microscope objectives are very demanding, as can be seen from the notes on the left of the imaging system box in figure 16.1. Optical glasses and their defects were discussed in chapter 5. Centering refers to the procedure for ensuring that all the centers of the surfaces of a nominally axisymmetric system do lie on one straight line to some tolerance; in simple systems this is done by grinding the edge of each lens component concentric with the axis of the component, this axis being the line joining the centers of its two surfaces. But it turns out that this is the least accurate of all the processes in making a lens: each spherical or plane surface can be made true to a fraction of a wavelength, the center thickness can be got to within perhaps 0.1 mm with reasonable effort, which is adequate for many but not all systems, but it is impossible to edge the tiny components of microscope objectives accurately enough to mount them straight into the metalwork. This has led to the development of a method of mounting which is widely adopted, with suitable variations, for many different high-precision systems. The component, edged as accurately as may be, is epoxied into a metal mount called a cell as in figure 16.2. The cell is then mounted in a lathe with a special centering chuck with enough degrees of freedom, and the chuck is adjusted until the lens is running true according to a suitable optical check. Then a fine skim is taken off the metal, usually with a diamond-point tool, so that the outside of the cell is running true with the lens axis. Each component is treated in this way, and the whole is assembled in an accurately bored tube. The point of this procedure is that diamond turning on a good lathe can produce a better concentric fit than edge- grinding the lens components.

16.2 Mounting an optical component for precision centering. The component is epoxied into the metal cell, and the latter is afterward turned concentric with the optical axis of the component.

Environmental aspects are relevant in, e.g., systems to go in aircraft or space vehicles, where violent and rapid temperature changes may occur. The effects of both thermal expansion and change of refractive index contribute, and a means must be found to correct the resulting focus shifts. This may be either passive, i.e., based simply on measuring the temperature (or other known cause of focal shift), or active, when an optoelectronic system is used to monitor the focus and apply corrections. An example of the latter is found in compact disk players, which contain a system like a simple microscope objective of NA about 0.4 which focuses a laser spot onto the disk and collects the light scattered back. Since the disks are merely stamped plastic, they may be out of plane by several tenths of a millimeter and the reading head incorporates a loudspeaker type of coil to adjust the focus in response to an error signal from part of the returned laser beam.

Other environmental hazards may be excessive humidity and contaminants carried to the lens surfaces as aerosols. It is usual in an optical design intended for use in such unfriendly environments to make the outer components of a resistant glass of hard crown or borosilicate crown type and arrange for the inside of the system to be sealed. One extreme example of an unfriendly environment is that endured by ordinary spectacles, and here a special *spectacle crown* of good resistance to all kinds of maltreatment is used.

Returning to figure 16.1, *metrical precision* refers to another aspect of accurate construction, the requirement that the distortion characteristics of the constructed imaging system be as in the design for, e.g., an objective used for aerial or satellite survey work or for mapmaking. In particular, the distortion function must be symmetrical about some point in the image plane designated as the intersection with the lens axis; otherwise, the calibration involves two coordinates rather than one. This is achieved by, among other things, giving great attention to centering, as described above.

Noise in detectors was discussed at some length in chapter 12. The characteristics on the right of the detector box in figure 16.1 depend on the kind of object and on the nature of the information to be extracted from the observation. Thus linearity is not so important if only a binary image is to be formed, but on the other hand it is important if an image with gray levels is required. We saw above how spatial resolution is matched to magnification. An obvious example where response time matters is where a changing phenomenon is being observed. We saw in chapter 12 how integration occurs in the photographic emulsion and in the photon-counting method of observation. Integration can also be regarded as a process of averaging to reduce the effect of noise, and this is done in *frame stores*; systems which will collect and store a complete TV frame and, in the more elaborate versions, accumulate repeat frames of the same picture and average them. This is a useful facility in conditions of low illumination.

## 16.3 Relay Optics and Field Lenses

The requirement to transport light beams or images over long distances can be met in many ways; for example, images can be transported as video signals over wires, radio, or single-mode optical fibers, and fiber bundles can also transport an image. In this section we discuss the uses of imaging systems in relay mode for this purpose.

A single-mode laser beam in free space expands as it propagates, as explained in chapter 9. One way to reduce this effect is to expand the beam first with a beam expander (figure 9.2). Then, if the waist of the expanded beam has the width $\omega_0$, the semiangle at which it expands is given by equation (9.5) and this can be made very small by making $\omega_0$ large enough. (However, in many cases the beam will be degraded and expanded by atmospheric turbulence effects if this approach is taken to extremes.) Another approach is to use a series of relay lenses as in figure 16.3 to reimage the beam waist. If the lenses are "thin" and if they all have focal length $f$, then the spacing between lenses is approximately $4f$; the exact expression is given in the figure. Provided the lens diameters are large enough to avoid appreciable truncation of the Gaussian profile, this method provides in principle transport over an indefinitely large distance.

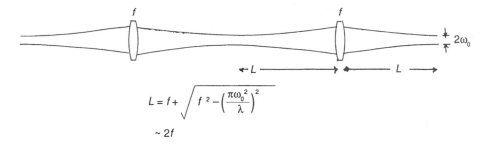

16.3 Relay lens system for a TEM $_{00}$ laser beam.

An image can be transported in the same way as the Gaussian beam, but a slight complication arises which can be understood by reference to figure 16.4, which is to be taken as within the geometrical optics model. In figure 16.4a, lens 1 images the extended object $O$ at $O'$ with magnification $-1$ and the principal ray from the extreme point of $O$ goes through the center of lens 1. Now, if this were to be followed by lens 2 at distance $2f$ from the intermediate image $O'$ (as in figure 16.3), it can be seen that the principal ray would not go through the center of lens 2 and in fact this lens must be made larger than lens 1 to avoid vignetting the beams from the edge of the original object. This difficulty is circumvented by means of a *field lens*, as in figure 16.4b. This field lens forms an image of lens 1 at lens 2 and thus returns the principal ray to the center of lens 2 and avoids vignetting. Quite long systems of this kind with several relay stages are used in remote viewing devices for radioactive areas and other inaccessible places and for submarine periscopes, but there are limits to the length that can be used in any particular case because of the buildup of aberrations. The long sequence of positive power lenses adds up to give large field curvature and astigmatism, and thus the usable field of view is restricted.

## 16.4 Pulsed Lasers in Optical Systems

We have already discussed in chapter 8 some points that arise in using cw lasers as sources for imaging systems. Pulsed lasers introduce further problems, not the least of which is the problem of damage to optical components. Bulk material damage takes the form of dielectric breakdown due

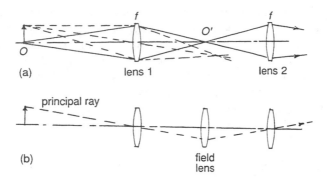

16.4 Relay lens system for a periscope.

to the high electric field in the light wave and the field may have been enhanced before breakdown by self-focusing due to nonlinear effects. Other damage mechanisms include heating due to absorption by local impurities (e.g., specks of unmelted batch from the glass furnace) and surface cracks or scratches. Depending on the type of laser, pulses lasting a few nanoseconds with energy of a few joules per pulse can be made, i.e., peak power of order $10^9$ watts. Damage thresholds to optical components are notoriously unrepeatable and vary with the type of optical glass, but it is generally thought that for short pulses peak energy densities inside the material above 1 J cm$^{-2}$ are dangerous. The threshold for surface damage is lower, depending very much on the quality of surface polish, and for most coatings it is lower still (Wood 1986 gives a good recent survey with numerical data).

One of the mechanisms of damage to thin films is of interest for its own sake. In chapter 10 we saw how to compute the reflection and transmission of a given multilayer structure. The intermediate matrix products obtained in the course of the computation can yield the electric and magnetic field strengths in standing waves in each layer corresponding to a given incident light intensity, and it can happen that some of these internal fields are much greater than that of the incident light, thus leading to a lower breakdown threshold than would otherwise be expected. It is therefore necessary when specifying coatings to warn the coating company about the expected pulse energy density.

In a multiple-lens system reflections off refracting surfaces can focus to a caustic, or strongly aberrated, focal spot and the energy density in this caustic can reach breakdown threshold even if the surfaces have been antireflection coated. If the surfaces have been left uncoated, for whatever reason, it may be desirable to trace rays through several reflections to check that no caustics are formed inside the glass.

A laser pulse may be multitransverse mode; i.e. it may in effect consist of several mutually incoherent modes traveling in different directions within a few milliradians.  The available information is rather sparse, but it is generally thought that in excimer lasers the beam spread or divergence may be a few milliradians, and although the beam as it leaves the output window is about 10 mm by 20 mm, the lateral coherence distance is only 2–3 mm inside this area; i.e., there is appreciable coherence only between points less than this distance apart.

## 16.5 Putting Together Optical Systems

Since the late 1970s the task of putting together a trial system or a system for a few special measurements has been made much easier by the emergence of comprehensive suppliers with detailed catalogs of optical and mechanical components.  Singlet lenses both spherical and cylindrical and achromatic doublets are available with focal lengths ranging from a few millimeters to a meter or so, usually in optical glass but sometimes also in fused silica (for the ultraviolet) and less frequently in infrared transmitting materials.  Different qualities of antireflection coating are offered, and full dimensional and surface finish tolerances are given, although the tolerances may not necessarily fit one's particular requirements.  Also, wide ranges of prisms, beam-splitters, interference filters, etc., are listed.  Some microscope objectives and photographic objectives are sometimes listed, although for these the specialized manufacturers are usually a better resource.  Component holders, fine adjustment mechanisms, and optical rails or benches are available, but often those of one manufacturer will not fit the optical components of another without modification.  Thus much can be done with off-the-shelf supplies, and for an initial experiment one should often try such products before going to custom optics manufacturers.

Custom manufacturing is needed often enough: to give a few examples, for exotic optical or mechanical materials, for special tolerances on surface finish or flatness, for suitability to unfriendly environments (e.g., temperature ranges), and for special aberrational tolerances.

Systems are often put together on special optical antivibration tables, which can have a metal surface with tapped holes at intervals or can be made of granite or a similar material.  Granite is said to be better for damping out residual vibrations, but metal tables provide better surfaces for clamping on components; in particular, if they are steel-topped, engineers' magnetic blocks can be used to facilitate rapid changes in an experimental layout.  Best of all is a cast iron engineer's marking out table or surface table; these do not distort when components are bolted down, and they can be mounted on inflated inner tubes for vibration insulation.

For the initial lining up of a system a HeNe laser is almost indispensable. The best procedure is to get the initial and the last components in position at approximately the right separation and at the same height above the datum. The HeNe laser is then used to carry this datum between them, and as each component is inserted its height and lateral position can be adjusted to hold the overall centration. At this stage residual surface reflections can help get components square to the optical axis if the system does have an axis of symmetry. However, as we have already noted in chapter 7, residual reflections in a system with laser illumination can give unwelcome interference effects in the direct beam, and if this matters for the purpose of the system, it is essential to get the antireflection coatings as efficient as possible. Then a good idea is to do the preliminary alignment with a different wavelength for which the coatings are not so efficient, so as to be able to use the back reflections.

The building up of dimensional tolerances in off-the-shelf components can lead to a final system in which the Gaussian properties, magnification, etc., are a long way from what was planned. It is not always practicable to allow for this in advance by computing the effects over all the tolerance ranges, as there are many possible combinations in a multicomponent system, and one cannot cope with the problem as in electronics with a few strategically placed potentiometers. The safest solution if the Gaussian properties are critical and if off-the-shelf components must be used is to split one or two lenses into two (e.g., a 100 mm focal length lens becomes two 200 mm focal length lenses) so that some adjustment is available by varying the spacing between them. This would not, of course, be necessary in a custom designed system which was properly toleranced.

# References

Birch, K. G., and Green, F. J. 1972, *Journal of Physics D*, 5, 1982–92. A detailed description of the techniques for generating holograms by computer.

Born, M., and Wolf, E. 1959, *Principles of Optics* (Elmsford, NY: Pergamon). This is the classic text on the theory of optics based on Maxwell's equations of the electromagnetic field; it is *not* an experimental text.

Dainty, J. C., ed. 1984, *Laser Speckle and Related Phenomena*, 2d ed. (Heidelberg: Springer). Chapters on the statistics of speckle formed under a variety of conditions, on methods of speckle reduction, and on applications of speckle.

Dobrowolski, J. A. 1978, section 8, "Coatings and Filters" in W. G. Driscoll and W. Vaughan, eds., *Handbook of Optics* (New York: McGraw-Hill). A comprehensive review of coatings and filters with graphical data for all types of filter.

Françon, M., and Mallick, S. 1971, *Polarization Interferometers* (New York: Wiley Interscience). Applications to testing and to microscopy.

Gaskill, J. D. 1978, *Linear Systems, Fourier Transforms and Optics* (New York: Wiley). A useful text on what is often called Fourier optics, covering the description of optical systems by Fourier transform methods and optical applications.

Gåsvik, K. J. 1987, *Optical Metrology* (New York: Wiley). A recent text on classical and holographic interferometry and other nondestructive testing methods)

Goodman, J. W. 1968, *Introduction to Fourier Optics* (New York: McGraw-Hill). A very well written general text on optical systems by Fourier methods.

Gottlieb, M., Ireland, C. L. M., and Ley, J. M. 1983, *Electro-optic and Acousto-optic Scanning and Deflection* (New York: Marcel Dekker). A specialist monograph with much practical detail.

Hariharan, P. 1984, *Optical Holography* (London: Cambridge University Press). A good general text with plenty of experimental detail about holography.

Jones, R. and Wykes, C. 1983, *Holographic and Speckle Interferometry* (New York: Cambridge University Press). A monograph aimed at non-destructive testing specialists as well as optical workers)

Kogelnik, H. and Li, T. 1966, *Applied Optics*, 5, 1550–1567. The classic reference on Gaussian laser beam theory.

Macleod, H. A. 1986, *Thin-Film Optical Filters* 2d ed. (Bristol: Adam Hilger) Probably the best reference on the theory and technique of multilayers.

Malacara, D., ed. 1978, *Optical Shop Testing* (New York: Wiley). A multiauthor text on workshop techniques of testing optical elements and systems.

Marshall, G. F., ed. 1985, *Laser beam scanning* (New York: Marcel Dekker). A specialist monograph with plenty of good experimental detail.

Michelson, A. A. 1902, *Light Waves and Their Uses* (Chicago: University of Chicago Press; a more recent edition was published in 1961, also by the University of Chicago Press). One of two classic texts by Michelson, valuable for its insights into his great experimental skills.

Palik, E. D. 1985, *Handbook of Optical Constants of Solids* (Academic Press). A collection of critically evaluated optical constants of nearly 40 metals, semiconductors, and dielectrics, covering wavelength ranges from belond the vacuum ultraviolet to the far-infrared.

Petit, R., ed. 1980, *Electromagnetic Theory of Gratings* (Berlin: Springer-Verlag). Powerful theoretical treatment of gratings according to exact electromagnetic theory; very mathematical.

Sheppard, C. 1984, *Theory and Practice of Optical Scanning Microscopy* (London: Academic Press). The first text on this relatively new method of optical microscopy.

Smith, H. M., ed. 1977, *Holographic Recording Materials* (Berlin: Springer-Verlag). Multiauthor treatment of a range of recording materials, not quite up-to-date but useful as far as it goes.

Smith, W. J. 1966, *Modern Optical Engineering* (New York: McGraw-Hill). A good general text on applied optics.

Stamnes, J. J. 1986, *Waves in Focal Regions* (Bristol: Adam Hilger). A research monograph on the structure of the region near the focus for scalar and electromagnetic waves; highly mathematical but plenty of useful graphical results.

Steel, W. H. 1983, *Interferometry* 2d ed. (Cambridge: Cambridge University Press). Probably still the best modern general text on interferometry.

Welford, W. T. 1964, *Progress in Optics*, IV, 241–280. Review article on aberration theory of diffraction gratings.

Welford, W. T. 1986, *Aberrations of Optical Systems* (Bristol: Adam Hilger). Aberration theory as required for the design of optical systems.

Wolfe, W. L. and Zissis, G. J. 1978, *The Infrared Handbook* (Washington, DC: Office of Naval Research). Useful collections of data on infrared materials and techniques, very compressed.

Wood, R. 1986, *Laser Damage in Optical Materials* (Bristol: Adam Hilger). Collections of numerical data which can be used as a basis for tolerancing in high-power laser systems.

# Index